Introduction to the Automatic Process Control

Introduction to the Automatic Process Control

Contributors :
Ricardo F. Escobar,
Manuel Adam-Medina, *et al.*

AURIS REFERENCE LTD.
London, UK

Introduction to the Automatic Process Control
Contributors : Ricardo F. Escobar, Manuel Adam-Medina, *et al.*

Auris Reference Ltd., UK

www.aurisreference.com

United Kingdom

Introduction to the Automatic Process Control

ISBN: 978-1-78154-505-8

British Library Cataloguing in Publication Data
A CIP record for this book is available from the British Library

Exclusively distributed by CBS Publishers & Distributors Pvt. Ltd.

Sales & Distribution Rights only for India, Pakistan, Bangladesh, Sri Lanka, Nepal and Bhutan.This book is not to be sold outside these territories.

PREFACE

An automatic control system is a preset closed-loop control system that requires no operator action. This assumes the process remains in the normal range for the control system. An automatic control system has two process variables associated with it: a controlled variable and a manipulated variable.

A controlled variable is the process variable that is maintained at a specified value or within a specified range. A manipulated variable is the process variable that is acted on by the control system to maintain the controlled variable at the specified value or within the specified range.

In the simplest type of an automatic control loop, a controller compares a measured value of a process with a desired set value, and processes the resulting error signal to change some input to the process, in such a way that the process stays at its set point despite disturbances.

This page left intentionally blank.

CONTENTS

Ricardo F. Escobar, Manuel Adam-Medina, Carlos D. García-Beltrán,
Víctor H. Olivares-Peregrino, David Juárez-Romero and Gerardo V. Guerrero-
Ramírez

This page left intentionally blank.

LIST OF CONTRIBUTORS

Ricardo F. Escobar

Centro Nacional de Investigación y Desarrollo Tecnológico, Tecnológico Nacional de México, Int. Internado Palmira S/N, Palmira, C.P. 62490 Cuernavaca, Morelos, Mexico; E-Mails: esjiri@cenidet.edu.mx (R.F.E.); adam@cenidet.edu.mx (M.A.-M.); cgarcia@cenidet.edu.mx (C.D.G.-B.); gerardog@cenidet.edu.mx (G.V.G.-R.)

Manuel Adam-Medina

Centro Nacional de Investigación y Desarrollo Tecnológico, Tecnológico Nacional de México, Int. Internado Palmira S/N, Palmira, C.P. 62490 Cuernavaca, Morelos, Mexico; E-Mails: esjiri@cenidet.edu.mx (R.F.E.); adam@cenidet.edu.mx (M.A.-M.); cgarcia@cenidet.edu.mx (C.D.G.-B.); gerardog@cenidet.edu.mx (G.V.G.-R.)

Carlos D. García-Beltrán

Centro Nacional de Investigación y Desarrollo Tecnológico, Tecnológico Nacional de México, Int. Internado Palmira S/N, Palmira, C.P. 62490 Cuernavaca, Morelos, Mexico; E-Mails: esjiri@cenidet.edu.mx (R.F.E.); adam@cenidet.edu.mx (M.A.-M.); cgarcia@cenidet.edu.mx (C.D.G.-B.); gerardog@cenidet.edu.mx (G.V.G.-R.)

Víctor H. Olivares-Peregrino

Centro Nacional de Investigación y Desarrollo Tecnológico, Tecnológico Nacional de México, Int. Internado Palmira S/N, Palmira, C.P. 62490 Cuernavaca, Morelos, Mexico; E-Mails: esjiri@cenidet.edu.mx (R.F.E.); adam@cenidet.edu.mx (M.A.-M.); cgarcia@cenidet.edu.mx (C.D.G.-B.); gerardog@cenidet.edu.mx (G.V.G.-R.)

David Juárez-Romero

Centro de Investigación en Ingeniería y Ciencias Aplicadas, Universidad Autónoma del Estado de Morelos, Av. Universidad 1001, Col. Chamilpa, C.P. 62209 Cuernavaca, Morelos, Mexico; E-Mail: djuarezr7@gmail.com

Gerardo V. Guerrero-Ramírez

Centro Nacional de Investigación y Desarrollo Tecnológico, Tecnológico Nacional de México, Int. Internado Palmira S/N, Palmira, C.P. 62490 Cuernavaca, Morelos, Mexico; E-Mails: esjiri@cenidet.edu.mx (R.F.E.); adam@cenidet.edu.mx (M.A.-M.); cgarcia@cenidet.edu.mx (C.D.G.-B.); gerardog@cenidet.edu.mx (G.V.G.-R.)

Chapter 1

Automatic Process Control

Automatic control is the application of control theory for regulation of processes without direct human intervention. In the simplest type of an automatic control loop, a controller compares a measured value of a process with a desired set value, and processes the resulting error signal to change some input to the process, in such a way that the process stays at its set point despite disturbances. This closed-loop control is an application of negative feedback to a system. The mathematical basis of control theory was begun in the 18th century, and advanced rapidly in the 20th.

Designing a system with features of automatic control generally requires the feeding of electrical or mechanical energy to enhance the dynamic features of an otherwise sluggish or variant, even errant system. The control is applied by regulating the energy feed.

Examples of Automatic Control

Automatic control can self-regulate a technical plant (such as a machine or an industrial process) operating condition or parameters by the controller with minimal human intervention. A regulator such as a thermostat is an example of a device studied in automatic control. Another possible example of Automatic Control are the ABS of a car.

Functions of Automatic control :

- Control
- Sensing
- Metrics
- Measurement
- Comparison
- Computation
- Correction

CONTROL SYSTEM

A **control system** is a device, or set of devices, that manages, commands, directs or regulates the behavior of other devices or systems. Industrial control systems are used in industrial production for controlling equipment or machines.

There are two common classes of control systems, open loop control systems and closed loop control systems. In open loop control systems output is generated based on inputs. In closed loop control systems current output is taken into consideration and corrections are made based on feedback. A closed loop system is also called a feedback control system. The human body is a classic example of feedback systems.

The term "control system" may be applied to the essentially manual controls that allow an operator, for example, to close and open a hydraulic press, perhaps including logic so that it cannot be moved unless safety guards are in place.

An automatic sequential control system may trigger a series of mechanical actuators in the correct sequence to perform a task. For example various electric and pneumatic transducers may fold and glue a cardboard box, fill it with product and then seal it in an automatic packaging machine. Programmable logic controllers are used in many cases such as this, but several alternative technologies exist.

In the case of linear feedback systems, a **control loop**, including sensors, control algorithms and actuators, is arranged in such a fashion as to try to regulate a variable at a setpoint or reference value. An example of this may increase the fuel supply to a furnace when a measured temperature drops. PID controllers are common and effective in cases such as this. Control systems that include some sensing of the results they are trying to achieve are making use of feedback and so can, to some extent, adapt to varying circumstances. Open-loop control systems do not make use of feedback, and run only in pre-arranged ways.

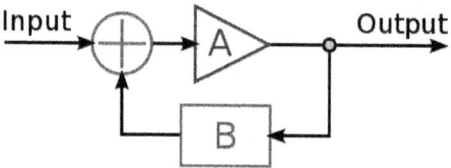

Fig. : A basic feedback loop.

Logic Control

Logic control systems for industrial and commercial machinery were historically implemented at mains voltage using interconnected relays, designed using ladder logic. Today, most such systems are constructed with programmable logic controllers (PLCs) or microcontrollers. The notation of ladder logic is still in use as a programming idiom for PLCs.

Logic controllers may respond to switches, light sensors, pressure switches, *etc.*, and can cause the machinery to start and stop various operations. Logic systems are used to sequence mechanical operations in many applications. PLC software can be written in many different ways – ladder diagrams, SFC – sequential function charts or in language terms known as statement lists.

Examples include elevators, washing machines and other systems with inter-related stop-go operations.

Logic systems are quite easy to design, and can handle very complex operations. Some aspects of logic system design make use of Boolean logic.

On–off Control

A thermostat is a simple negative feedback controller: when the temperature (the "process variable" or PV) goes below a set point (SP), the heater is switched on. Another example could be a pressure switch on an air compressor. When the pressure (PV) drops below the threshold (SP), the pump is powered. Refrigerators and vacuum pumps contain similar mechanisms operating in reverse, but still providing negative feedback to correct errors.

Simple on–off feedback control systems like these are cheap and effective. In some cases, like the simple compressor example, they may represent a good design choice.

In most applications of on–off feedback control, some consideration needs to be given to other costs, such as wear and tear of control valves and perhaps other start-up costs when power is reapplied each time the PV drops. Therefore, practical on–off control systems are designed to include hysteresis which acts as a deadband, a region around the setpoint value in which no control action occurs. The width of deadband may be adjustable or programmable.

Linear Control

Linear control systems use linear negative feedback to produce a control signal mathematically based on other variables, with a view to maintain the controlled process within an acceptable operating range.

The output from a linear control system into the controlled process may be in the form of a directly variable signal, such as a valve that may be 0 or 100% open or anywhere in between. Sometimes this is not feasible and so, after calculating the current required corrective signal, a linear control system may repeatedly switch an actuator, such as a pump, motor or heater, fully on and then fully off again, regulating the duty cycle using pulse-width modulation.

When controlling the temperature of an industrial furnace, it is usually better to control the opening of the fuel valve *in proportion to* the current needs of the furnace. This helps avoid thermal shocks and applies heat more effectively.

Proportional negative-feedback systems are based on the difference between the required set point (SP) and process value (PV). This difference is called the *error*. Power is applied in direct proportion to the current measured error, in the correct sense so as to tend to reduce the error and therefore avoid positive feedback. The amount of corrective action that is applied for a given error is set by the gain or sensitivity of the control system.

At low gains, only a small corrective action is applied when errors are detected. The system may be safe and stable, but may be sluggish in response to changing conditions. Errors will remain uncorrected for relatively long periods of time and the system is over-damped. If the proportional gain is increased, such systems become more responsive and errors are dealt with more quickly. There is an optimal value for the gain setting when the overall system is said to be critically damped. Increases in loop gain beyond this point lead to oscillations in the PV and such a system is under-damped.

In real systems, there are practical limits to the range of the manipulated variable (MV). For example, a heater can be off or fully on, or a valve can be closed or fully open. Adjustments to the gain simultaneously alter the range of error values over which the MV is between these limits. The width of this range, in units of the error variable and therefore of the PV, is called the *proportional band* (PB). While the gain is useful in mathematical treatments, the proportional band is often used in practical situations. They both refer to the same thing, but the PB has an inverse relationship to gain – higher gains result in narrower PBs, and *vice versa*.

Under-damped Furnace Example

In the furnace example, suppose the temperature is increasing towards a set point at which, say, 50% of the available power will be required for steady-state. At low temperatures, 100% of available power is applied. When the process value (PV) is within, say 10° of the SP the heat input begins to be reduced by the proportional controller (note that this implies a 20° proportional band (PB) from full to no power input, evenly spread around the setpoint value).

At the setpoint the controller will be applying 50% power as required, but stray stored heat within the heater sub-system and in the walls of the furnace will keep the measured temperature rising beyond what is required. At 10° above SP, we reach the top of the proportional band (PB) and no power is applied, but the temperature may continue to rise even further before beginning to fall back. Eventually as the PV falls back into the PB, heat is applied again, but now the heater and the furnace walls are too cool and the temperature falls too low before its fall is arrested, so that the oscillations continue.

Over-damped Furnace Example

The temperature oscillations that an under-damped furnace control system produces are unacceptable for many reasons, including the waste of fuel and time (each oscillation cycle may take many minutes), as well as the likelihood of seriously overheating both the furnace and its contents.

Suppose that the gain of the control system is reduced drastically and it is restarted. As the temperature approaches, say 30° below SP (60° proportional band (PB)), the heat input begins to be reduced, the rate of heating of the furnace has time to slow and, as the heat is still further reduced, it eventually is brought up to set point, just as 50% power input is reached and the furnace is operating as required. There was some wasted time while the furnace crept to its final temperature using only 52% then 51% of available power, but at least no harm was done.

By carefully increasing the gain (*i.e.*, reducing the width of the PB) this over-damped and sluggish behavior can be improved until the system is critically damped for this SP temperature. Doing this is known as 'tuning' the control system. A well-tuned proportional furnace temperature control system will usually be more effective than on-off control, but will still respond more slowly than the furnace could under skillful manual control.

SENSOR

A **sensor** is an object whose purpose is to detect events or changes in its environment, and then provide a corresponding output. A sensor is a type of transducer; sensors may provide various types of output, but typically use electrical or optical signals. For example, a thermocouple generates a known voltage (the output) in response to its temperature (the environment). A mercury-in-glass thermometer, similarly, converts measured temperature into expansion and contraction of a liquid, which can be read on a calibrated glass tube.

Sensors are used in everyday objects such as touch-sensitive elevator buttons (tactile sensor) and lamps which dim or brighten by touching the base, besides innumerable applications of which most people are never aware. With advances in micro machinery and easy-to-use micro controller platforms, the uses of sensors have expanded beyond the more traditional fields of temperature, pressure or flow measurement, for example into MARG sensors. Moreover, analog sensors such as potentiometers and force-sensing resistors are still widely used. Applications include manufacturing and machinery, airplanes and aerospace, cars, medicine and robotics.it is also included in our day-to-day life.

A sensor's sensitivity indicates how much the sensor's output changes when the input quantity being measured changes. For instance, if the mercury in a thermometer moves 1 cm when the temperature changes by 1°C, the sensitivity is 1 cm/°C (it is basically the slope Dy/Dx assuming a linear characteristic). Some sensors can also have an impact on what they measure; for instance, a room temperature thermometer inserted into a hot cup of liquid cools the liquid while the liquid heats the thermometer. Sensors need to be designed to have a small effect on what is measured; making the sensor smaller often improves this and may introduce other advantages. Technological progress allows more and more sensors to be manufactured on a microscopic scale as microsensors using MEMS technology. In most cases, a microsensor reaches a significantly higher speed and sensitivity compared with macroscopic approaches.

Chemical Sensor

A chemical sensor is a self-contained analytical device that can provide information about the chemical composition of its environment, that is, a liquid or a gas phase. The information is provided in the form of a measurable physical signal that is correlated with the concentration of a certain chemical species (termed as analyte). Two main steps are involved in the functioning of a chemical sensor, namely, recognition and transduction. In the recognition step, analyte molecules interact selectively with receptor molecules or sites included in the structure of the recognition element of the sensor.

Consequently, a characteristic physical parameter varies and this variation is reported by means of an integrated transducer that generates the output signal. A chemical sensor based on recognition material of biological nature is a biosensor. However, as synthetic biomimetic materials are going to substitute to some extent recognition biomaterials, a sharp distinction between a biosensor and a standard chemical sensor is superfluous. Typical biomimetic materials used in sensor development are molecularly imprinted polymers and aptamers.

Biosensor

In biomedicine and biotechnology, sensors which detect analytes thanks to a biological component, such as cells, protein, nucleic acid or biomimetic polymers, are called biosensors. Whereas a non-biological sensor, even organic (=carbon chemistry), for biological analytes is referred to as sensor or nanosensor (such a microcantilevers). This terminology applies for both in vitro and in vivo applications.

The encapsulation of the biological component in biosensors, presents a slightly different problem that ordinary sensors; this can either be done by means of a semipermeable barrier, such as a dialysis membrane or a hydrogel, or a 3D polymer matrix, which either physically constrains the sensing macromolecule or chemically constrains the macromolecule by bounding it to the scaffold.

Performance Metric

A **performance metric** is that which determines an organization's behavior and performance. Performance metrics measure an organization's activities and performance. It should support a range of stakeholder needs from customers, shareholders to employees. While traditionally many metrics are finance based, inwardly focusing on the performance of the organization, metrics may also focus on the performance against customer requirements and value.

In project management, performance metrics are used to assess the health of the project and consist of the measuring of seven criteria: safety, time, cost, resources, scope, quality, and actions. In call centres, performance metrics help capture internal performance and can include productivity measurements and the quality of service provided by the customer service advisor. These metrics can include: Calls Answered, Calls Abandoned, Average Handle Time and Average Wait Time.

Developing performance metrics usually follows a process of:

1. Establishing critical processes/customer requirements
2. Identifying specific, quantifiable outputs of work
3. Establishing targets against which results can be scored

A criticism of performance metrics is that when the value of information is computed using mathematical methods, it shows that even performance metrics professionals choose measures that have little value. This is referred to as the "measurement inversion". For example, metrics seem to emphasize what organizations find immediately measurable – even if those are low value – and tend to ignore high value measurements simply because they seem harder to measure (whether they are or not).

To correct for the measurement inversion other methods, like applied information economics, introduce the "value of information analysis" step in the process so that metrics focus on high-value measures. Organizations where this has been applied find that they define completely different metrics than they otherwise would have and, often, fewer metrics. For projects, the effort to collect a metric has to be weighed against its value as projects are temporary endeavors performed with finite resources.

There are a variety of ways in which organizations may react to results. This may be to trigger specific activity relating to performance (*i.e.*, an improvement plan) or to use the data merely for statistical information. Often closely tied in with outputs, performance metrics should usually encourage improvement, effectiveness and appropriate levels of control.

Performance metrics are often linked in with corporate strategy and are often derived in order to measure performance against a critical success factor.

COMPUTATION

Computation is any type of calculation that follows a well-defined model understood and expressed as, for example, an algorithm, or a protocol.

The study of computation is paramount to the discipline of computer science.

Classification

Computation can be classified by mainly three unique criteria: digital versus analog, sequential versus parallel versus concurrent, batch versus interactive.

In practice, digital computation aids simulation of natural processes (for example, evolutionary computation), including those that are naturally described by analog models of computation (for example, artificial neural network).

Comparison to Calculation

Calculation is a term for the computation of numbers, while computation is a wider reaching term for information processing in general.

Physical Phenomenon

A computation can be seen as a purely physical phenomenon occurring inside a closed physical system called a computer. Examples of such physical systems include digital computers, mechanical computers, quantum computers, DNA computers, molecular computers, analog computers or wetware computers. This point of view is the one adopted by the branch of theoretical physics called the physics of computation.

Mathematical Models

In the theory of computation, a diversity of mathematical models of computers have been developed. Typical mathematical models of computers are the following:

- State models including Turing machine, push-down automaton, finite state automaton, and PRAM
- Functional models including lambda calculus
- Logical models including logic programming
- Concurrent models including actor model and process calculi

MEASUREMENT

Measurement is the assignment of a number to a characteristic of an object or event, which can be compared with other objects or events. The scope and application of a measurement is dependent on the context and discipline. In the natural sciences and engineering, measurements do not apply to nominal properties of objects or events, which is consistent with the guidelines of the *International vocabulary of metrology* published by the International Bureau of Weights and Measures. However, in other fields such as statistics as well as the social and behavioral sciences, measurements can have multiple levels, which would include nominal, ordinal, interval, and ratio scales.

Measurement is a cornerstone of trade, science, technology, and quantitative research in many disciplines. Historically, many measurement systems existed for the varied fields of human existence to facilitate comparisons in these fields. Often these were achieved by local agreements between trading partners or collaborators. Since the 18th century, developments progressed towards unifying, widely accepted standards that resulted in the modern International System of Units (SI). This system reduces all physical measurements to a mathematical combination of seven base units. The science of measurement is pursued in the field of metrology.

Methodology

The measurement of a property may be categorized by the following criteria: type, magnitude, unit, and uncertainty. They enable unambiguous comparisons between measurements.

The *type* or *level* of measurement is a taxonomy for the methodological character of a comparison. For example, two states of a property may be compared by ratio, difference, or ordinal preference. The type is commonly not explicitly expressed, but implicit in the definition of a measurement procedure. The *magnitude* is the numerical value of the characterization, usually obtained with a suitably chosen measuring instrument.

A *unit* assigns a mathematical weighting factor to the magnitude that is derived as a ratio to the property of an artifact used as standard or a natural physical quantity. An *uncertainty* represents the random and systemic errors of the measurement procedure; it indicates a confidence level in the measurement. Errors are evaluated by methodically repeating measurements and considering the accuracy and precision of the measuring instrument.

Standardization of Measurement Units

Measurements most commonly use the International System of Units (SI) as a comparison framework. The system defines seven fundamental units: kilogram, metre, candela, second, ampere, kelvin, and mole. Six of these units are defined without reference to a particular physical object which serves as a standard (artifact-free), with the exception of the kilogram which is still embodied in an artifact which rests at the BIPM outside Paris.

Artifact-free definitions fix measurements at an exact value related to a physical constant or other invariable phenomena in nature, in contrast to standard artifacts which are subject to deterioration or destruction. Instead, the measurement unit can only ever change through increased accuracy in determining the value of the constant it is tied to.

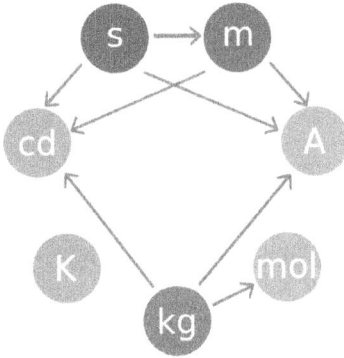

Fig. : The seven base units in the SI system. Arrows point from units to those that depend on them.

TYPES OF PROCESSES USING PROCESS CONTROL

Process control is an engineering discipline that deals with architectures, mechanisms and algorithms for maintaining the output of a specific process within a desired range. For instance, the temperature of a chemical reactor may be controlled to maintain a consistent product output.

Process control is extensively used in industry and enables mass production of consistent products from continuously operated processes such as oil refining, paper manufacturing, chemicals, power plants and many others. Process control enables automation, by which a small staff of operating personnel can operate a complex process from a central control room.

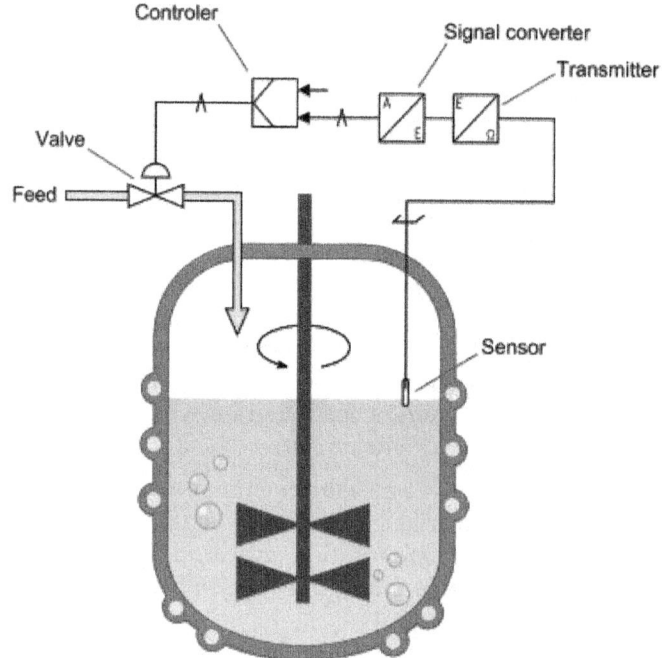

Fig. : Example of control system of a continuous stirred-tank reactor.

Process control may either use feedback or it may be open loop. Control may also be continuous (automobile cruise control) or cause a sequence of discrete events, such as a timer on a lawn sprinkler (on/off) or controls on an elevator (logical sequence).

A thermostat on a heater is an example of control that is on or off. A temperature sensor turns the heat source on if the temperature falls below the set point and turns the heat source off when the set point is reached. There is no measurement of the difference between the set point and the measured temperature (*e.g.* no error measurement) and no adjustment to the rate at which heat is added other than all or none.

A familiar example of feedback control is cruise control on an automobile. Here speed is the **measured variable**. The operator (driver) adjusts the desired speed **set point** (*e.g.* 100 km/hr) and the controller monitors the speed sensor and compares the measured speed to the set point. Any deviations, such as changes in grade, drag, wind speed or even using a different grade of fuel (for example an ethanol blend) are corrected by the controller making a compensating adjustment to the fuel valve open position, which is the **manipulated variable**. The controller makes adjustments having information only about the error (magnitude, rate of change or cumulative error) although settings known as *tuning* are used to achieve stable control. The operation of such controllers is the subject of control theory.

A commonly used control device called a programmable logic controller, or a PLC, is used to read a set of digital and analog inputs, apply a set of logic statements, and generate a set of analog and digital outputs.

For example, if an adjustable valve were used to hold level in a tank the logical statements would compare the equivalent pressure at depth setpoint to the pressure reading of a sensor below the normal low liquid level and determine whether more or less valve opening was necessary to keep the level constant. A PLC output would then calculate an incremental amount of change in the valve position. Larger more complex systems can be controlled by a Distributed Control System (DCS) or SCADA system.

AUTOMATION

Automation or *automatic control*, is the use of various control systems for operating equipment such as machinery, processes in factories, boilers and heat treating ovens, switching on telephone networks, steering and stabilization of ships, aircraft and other applications with minimal or reduced human intervention. Some processes have been completely automated.

The biggest benefit of automation is that it saves labor, however, it is also used to save energy and materials and to improve quality, accuracy and precision.

The term *automation*, inspired by the earlier word *automatic* (coming from *automaton*), was not widely used before 1947, when General Motors established the automation department. It was during this time that industry was rapidly adopting feedback controllers, which were introduced in the 1930s.

Automation has been achieved by various means including mechanical, hydraulic, pneumatic, electrical, electronic devices and computers, usually in combination. Complicated systems, such as modern factories, airplanes and ships typically use all these combined techniques.

Types of Automation

1. Discrete Control (on/off)

One of the simplest types of control is *on-off* control. An example is the thermostats used on household appliances. Electromechanical thermostats used in

HVAC may only have provision for on/off control of heating or cooling systems. Electronic controllers may add multiple stages of heating and variable fan speed control.

Sequence control, in which a programmed sequence of *discrete* operations is performed, often based on system logic that involves system states. An elevator control system is an example of sequence control.

2. Continuous Control

The advanced type of automation that revolutionized manufacturing, aircraft, communications and other industries, is feedback control, which is usually *continuous* and involves taking measurements using a sensor and making calculated adjustments to keep the measured variable within a set range.

Open and Closed Loop

All the elements constituting the measurement and control of a single variable are called a *control loop*. Control that uses a measured signal, feeds the signal back and compares it to a set point, calculates and sends a return signal to make a correction, is called *closed loop* control. If the controller does not incorporate feedback to make a correction then it is *open loop*.

Loop control is normally accomplished with a controller. The theoretical basis of open and closed loop automation is control theory.

Sequential Control and Logical Sequence or System State Control

Sequential control may be either to a fixed sequence or to a logical one that will perform different actions depending on various system states. An example of an adjustable but otherwise fixed sequence is a timer on a lawn sprinkler.

States refer to the various conditions that can occur in a use or sequence scenario of the system. An example is an elevator, which uses logic based on the system state to perform certain actions in response to its state and operator input. For example, if the operator presses the floor n button, the system will respond depending on whether the elevator is stopped or moving, going up or down, or if the door is open or closed, and other conditions.

An early development of sequential control was relay logic, by which electrical relays engage electrical contacts which either start or interrupt power to a device. Relays were first used in telegraph networks before being developed for controlling other devices, such as when starting and stopping industrial-sized electric motors or opening and closing solenoid valves. Using relays for control purposes allowed event-driven control, where actions could be triggered out of sequence, in response to external events. These were more flexible in their response than the rigid single-sequence cam timers. More complicated examples involved maintaining safe sequences for devices such as swing bridge controls, where a lock bolt needed to be disengaged before the bridge could be moved, and the lock bolt could not be released until the safety gates had already been closed.

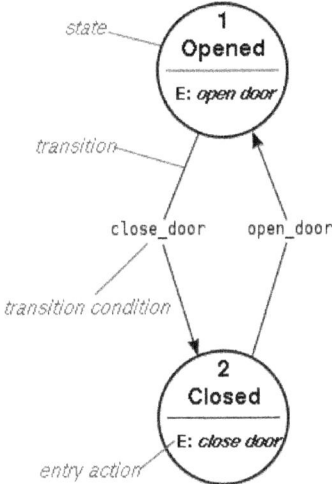

Fig. : This state diagram shows how UML can be used for designing a door system that can only be opened and closed.

The total number of relays, cam timers and drum sequencers can number into the hundreds or even thousands in some factories. Early programming techniques and languages were needed to make such systems manageable, one of the first being ladder logic, where diagrams of the interconnected relays resembled the rungs of a ladder. Special computers called programmable logic controllers were later designed to replace these collections of hardware with a single, more easily re-programmed unit.

In a typical hard wired motor start and stop circuit (called a *control circuit*) a motor is started by pushing a "Start" or "Run" button that activates a pair of electrical relays. The "lock-in" relay locks in contacts that keep the control circuit energized when the push button is released. (The start button is a normally open contact and the stop button is normally closed contact.) Another relay energizes a switch that powers the device that throws the motor starter switch (three sets of contacts for three phase industrial power) in the main power circuit. Large motors use high voltage and experience high in-rush current, making speed important in making and breaking contact. This can be dangerous for personnel and property with manual switches. All contacts are held engaged by their respective electromagnets until a "stop" or "off" button is pressed, which de-energizes the lock in relay.

Commonly interlocks are added to a control circuit. Suppose that the motor in the example is powering machinery that has a critical need for lubrication. In this case an interlock could be added to insure that the oil pump is running before the motor starts. Timers, limit switches and electric eyes are other common elements in control circuits.

Solenoid valves are widely used on compressed air or hydraulic fluid for powering actuators on mechanical components. While motors are used to supply continuous rotary motion, actuators are typically a better choice for intermittently creating a limited range of movement for a mechanical component, such as moving various mechanical arms, opening or closing valves, raising heavy press rolls, applying pressure to presses.

Computer Control

Computers can perform both sequential control and feedback control, and typically a single computer will do both in an industrial application. Programmable logic controllers (PLCs) are a type of special purpose microprocessor that replaced many hardware components such as timers and drum sequencers used in relay logic type systems. General purpose process control computers have increasingly replaced stand alone controllers, with a single computer able to perform the operations of hundreds of controllers.

Process control computers can process data from a network of PLCs, instruments and controllers in order to implement typical (such as PID) control of many individual variables or, in some cases, to implement complex control algorithms using multiple inputs and mathematical manipulations. They can also analyze data and create real time graphical displays for operators and run reports for operators, engineers and management.

Control of an automated teller machine (ATM) is an example of an interactive process in which a computer will perform a logic derived response to a user selection based on information retrieved from a networked database. The ATM process has similarities with other online transaction processes. The different logical responses are called *scenarios*. Such processes are typically designed with the aid of use cases and flowcharts, which guide the writing of the software code.

History

The earliest feedback control mechanism was used to tent the sails of windmills. It was patented by Edmund Lee in 1745.

The centrifugal governor, which dates to the last quarter of the 18th century, was used to adjust the gap between millstones. The centrifugal governor was also used in the automatic flour mill developed by Oliver Evans in 1785, making it the first completely automated industrial process. The governor was adopted by James Watt for use on a steam engine in 1788 after Watt's partner Boulton saw one at a flour mill Boulton & Watt were building.

The governor could not actually hold a set speed; the engine would assume a new constant speed in response to load changes. The governor was able to handle smaller variations such as those caused by fluctuating heat load to the boiler. Also, there was a tendency for oscillation whenever there was a speed change.

As a consequence, engines equipped with this governor were not suitable for operations requiring constant speed, such as cotton spinning.

Several improvements to the governor, plus improvements to valve cut-off timing on the steam engine, made the engine suitable for most industrial uses before the end of the 19th century. Advances in the steam engine stayed well ahead of science, both thermodynamics and control theory.

The governor received relatively little scientific attention until James Clerk Maxwell published a paper that established the beginning of a theoretical basis for understanding control theory. Development of the electronic amplifier during the 1920s, which was important for long distance telephony, required a higher signal to noise ratio, which was solved by negative feedback noise cancellation. This and other telephony applications contributed to control theory. Military applications during the Second World War that contributed to and benefited from control theory were fire-control systems and aircraft controls. The word "automation" itself was coined in the 1940s by General Electric. The so-called classical theoretical treatment of control theory dates to the 1940s and 1950s.

Relay logic was introduced with factory electrification, which underwent rapid adaption from 1900 though the 1920s. Central electric power stations were also undergoing rapid growth and operation of new high pressure boilers, steam turbines and electrical substations created a large demand for instruments and controls.

Central control rooms became common in the 1920s, but as late as the early 1930s, most process control was on-off. Operators typically monitored charts drawn by recorders that plotted data from instruments. To make corrections, operators manually opened or closed valves or turned switches on or off. Control rooms also used color coded lights to send signals to workers in the plant to manually make certain changes.

Controllers, which were able to make calculated changes in response to deviations from a set point rather than on-off control, began being introduced the 1930s. Controllers allowed manufacturing to continue showing productivity gains to offset the declining influence of factory electrification.

In 1959 Texaco's Port Arthur refinery became the first chemical plant to use digital control. Conversion of factories to digital control began to spread rapidly in the 1970s as the price of computer hardware fell.

SIGNIFICANT APPLICATIONS

The automatic telephone switchboard was introduced in 1892 along with dial telephones. By 1929, 31.9% of the Bell system was automatic. Automatic telephone switching originally used vacuum tube amplifiers and electro-mechanical switches, which consumed a large amount of electricity. Call volume eventually grew so fast that it was feared the telephone system would consume all electricity production, prompting Bell Labs to begin research on the transistor.

The logic performed by telephone switching relays was the inspiration for the digital computer.

The first commercially successful glass bottle blowing machine was an automatic model introduced in 1905. The machine, operated by a two-man crew working 12-hour shifts, could produce 17,280 bottles in 24 hours, compared to 2,880 bottles made by a crew of six men and boys working in a shop for a day. The cost of making bottles by machine was 10 to 12 cents per gross compared to $1.80 per gross by the manual glassblowers and helpers.

Sectional electric drives were developed using control theory. Sectional electric drives are used on different sections of a machine where a precise differential must be maintained between the sections. In steel rolling, the metal elongates as it passes through pairs of rollers, which must run at successively faster speeds. In paper making the paper sheet shrinks as it passes around steam heated drying arranged in groups, which must run at successively slower speeds. The first application of a sectional electric drive was on a paper machine in 1919. One of the most important developments in the steel industry during the 20th century was continuous wide strip rolling, developed by Armco in 1928.

Before automation many chemicals were made in batches. In 1930, with the widespread use of instruments and the emerging use of controllers, the founder of Dow Chemical Co. was advocating continuous production.

Self-acting machine tools that displaced hand dexterity so they could be operated by boys and unskilled laborers were developed by James Nasmyth in the 1840s. Machine tools were automated with Numerical control (NC) using punched paper tape in the 1950s. This soon evolved into computerized numerical control (CNC).

Today extensive automation is practiced in practically every type of manufacturing and assembly process. Some of the larger processes include electrical power generation, oil refining, chemicals, steel mills, plastics, cement plants, fertilizer plants, pulp and paper mills, automobile and truck assembly, aircraft production, glass manufacturing, natural gas separation plants, food and beverage processing, canning and bottling and manufacture of various kinds of parts. Robots are especially useful in hazardous applications like automobile spray painting. Robots are also used to assemble electronic circuit boards. Automotive welding is done with robots and automatic welders are used in applications like pipelines.

Advantages and Disadvantages

The main advantages of automation are:
- Increased throughput or productivity.
- Improved quality or increased predictability of quality.
- Improved robustness (consistency), of processes or product.
- Increased consistency of output.
- Reduced direct human labor costs and expenses.

The following methods are often employed to improve productivity, quality, or robustness.

- Install automation in operations to reduce cycle time.
- Install automation where a high degree of accuracy is required.
- Replacing human operators in tasks that involve hard physical or monotonous work.
- Replacing humans in tasks done in dangerous environments (*i.e.* fire, space, volcanoes, nuclear facilities, underwater, *etc.*)
- Performing tasks that are beyond human capabilities of size, weight, speed, endurance, *etc.*
- Economic improvement: Automation may improve in economy of enterprises, society or most of humanity. For example, when an enterprise invests in automation, technology recovers its investment; or when a state or country increases its income due to automation like Germany or Japan in the 20th Century.
- Reduces operation time and work handling time significantly.
- Frees up workers to take on other roles.
- Provides higher level jobs in the development, deployment, maintenance and running of the automated processes.

The main disadvantages of automation are:

- Security Threats/Vulnerability: An automated system may have a limited level of intelligence, and is therefore more susceptible to committing errors outside of its immediate scope of knowledge (*e.g.*, it is typically unable to apply the rules of simple logic to general propositions).
- Unpredictable/excessive development costs: The research and development cost of automating a process may exceed the cost saved by the automation itself.
- High initial cost: The automation of a new product or plant typically requires a very large initial investment in comparison with the unit cost of the product, although the cost of automation may be spread among many products and over time.

In manufacturing, the purpose of automation has shifted to issues broader than productivity, cost, and time.

Lights out Manufacturing

Lights out manufacturing is when a production system is 100% or near to 100% automated (not hiring any workers). In order to eliminate the need for labor costs all together.

Health and Environment

The costs of automation to the environment are different depending on the technology, product or engine automated. There are automated engines that consume more energy resources from the Earth in comparison with previous engines and those that do the opposite too. Hazardous operations, such as oil refining, the manufacturing of industrial chemicals, and all forms of metal working, were always early contenders for automation.

Convertibility and Turnaround Time

Another major shift in automation is the increased demand for flexibility and convertibility in manufacturing processes. Manufacturers are increasingly demanding the ability to easily switch from manufacturing Product A to manufacturing Product B without having to completely rebuild the production lines. Flexibility and distributed processes have led to the introduction of Automated Guided Vehicles with Natural Features Navigation.

Digital electronics helped too. Former analogue-based instrumentation was replaced by digital equivalents which can be more accurate and flexible, and offer greater scope for more sophisticated configuration, parametrization and operation. This was accompanied by the fieldbus revolution which provided a networked (*i.e.* a single cable) means of communicating between control systems and field level instrumentation, eliminating hard-wiring.

Discrete manufacturing plants adopted these technologies fast. The more conservative process industries with their longer plant life cycles have been slower to adopt and analogue-based measurement and control still dominates. The growing use of Industrial Ethernet on the factory floor is pushing these trends still further, enabling manufacturing plants to be integrated more tightly within the enterprise, via the internet if necessary. Global competition has also increased demand for Reconfigurable Manufacturing Systems.

Automation Tools

Engineers can now have numerical control over automated devices. The result has been a rapidly expanding range of applications and human activities. Computer-aided technologies (or CAx) now serve as the basis for mathematical and organizational tools used to create complex systems. Notable examples of CAx include Computer-aided design (CAD software) and Computer-aided manufacturing (CAM software). The improved design, analysis, and manufacture of products enabled by CAx has been beneficial for industry.

Information technology, together with industrial machinery and processes, can assist in the design, implementation, and monitoring of control systems. One example of an industrial control system is a programmable logic controller (PLC). PLCs are specialized hardened computers which are frequently used to synchronize the flow of inputs from (physical) sensors and events with the flow of outputs to actuators and events.

Human-machine interfaces (HMI) or computer human interfaces (CHI), formerly known as *man-machine interfaces*, are usually employed to communicate with PLCs and other computers. Service personnel who monitor and control through HMIs can be called by different names. In industrial process and manufacturing environments, they are called operators or something similar. In boiler houses and central utilities departments they are called stationary engineers.

Different types of automation tools exist:

- ANN - Artificial neural network
- DCS - Distributed Control System
- HMI - Human Machine Interface
- SCADA - Supervisory Control and Data Acquisition
- PLC - Programmable Logic Controller
- Instrumentation
- Motion control
- Robotics

When it comes to Factory Automation, Host Simulation Software (HSS) is a commonly used testing tool that is used to test the equipment software. HSS is used to test equipment performance with respect to Factory Automation standards (timeouts, response time, processing time).

Limitations to Automation

- Current technology is unable to automate all the desired tasks.
- Many operations using automation have large amounts of invested capital and produce high volumes of product, making malfunctions extremely costly and potentially hazardous. Therefore, some personnel are needed to insure that the entire system functions properly and that safety and product quality are maintained.
- As a process becomes increasingly automated, there is less and less labor to be saved or quality improvement to be gained. This is an example of both diminishing returns and the logistic function.
- As more and more processes become automated, there are fewer remaining non-automated processes. This is an example of exhaustion of opportunities. New technological paradigms may however set new limits that surpass the previous limits.

Current Limitations

Many roles for humans in industrial processes presently lie beyond the scope of automation. Human-level pattern recognition, language comprehension, and language production ability are well beyond the capabilities of modern mechanical and computer systems. Tasks requiring subjective assessment or synthesis of complex sensory data, such as scents and sounds, as well as high-level tasks such as strategic planning, currently require human expertise. In many cases, the use of humans is more cost-effective than mechanical approaches even where automation of industrial tasks is possible. Overcoming these obstacles is a theorized path to post-scarcity economics.

Paradox of Automation

The Paradox of Automation says that the more efficient the automated system, the more crucial the human contribution of the operators. Humans are less involved, but their involvement becomes more critical.

If an automated system has an error, it will multiply that error until it's fixed or shut down. This is where human operators come in.

A fatal example of this was Air France Flight 447, where a failure of automation put the pilots into a manual situation they were not prepared for.

RECENT AND EMERGING APPLICATIONS

Automated Retail

Food and Drink

The food retail industry has started to apply automation to the ordering process; McDonald's has introduced touch screen ordering and payment systems in many of its restaurants, reducing the need for as many cashier employees. The University of Texas at Austin has introduced fully automated cafe retail locations. Some Cafes and restaurants have utilized mobile and tablet "apps" to make the ordering process more efficient by customers ordering and paying on their device. Some restaurants have automated food delivery to customers tables using a Conveyor belt system. The use of robots is sometimes employed to replace waiting staff.

Stores

Many Supermarkets and even smaller stores are rapidly introducing Self checkout systems reducing the need for employing checkout workers.

Online shopping could be considered a form of automated retail as the payment and checkout are through an automated Online transaction processing system. Other forms of automation can also be an integral part of online shopping, for example the deployment of automated warehouse robotics such as that applied by Amazon using Kiva Systems.

Automated Mining

Involves the removal of human labor from the mining process. The mining industry is currently in the transition towards Automation. Currently it can still require a large amount of human capital, particularly in the third world where labor costs are low so there is less incentive for increasing efficiency through automation.

Automated Video Surveillance

The Defense Advanced Research Projects Agency (DARPA) started the research and development of automated visual surveillance and monitoring (VSAM) program, between 1997 and 1999, and airborne video surveillance (AVS) programs, from 1998 to 2002. Currently, there is a major effort underway in the vision community to develop a fully automated tracking surveillance system. Automated video surveillance monitors people and vehicles in real time within a busy environment. Existing automated surveillance systems are based on the environment they are primarily designed to observe, *i.e.*, indoor, outdoor or airborne, the amount of sensors that the automated system can handle and the mobility of sensor, *i.e.*, stationary camera vs. mobile camera. The purpose of a surveillance system is to record properties and trajectories of objects in a given area, generate warnings or notify designated authority in case of occurrence of particular events.

Automated Highway Systems

As demands for safety and mobility have grown and technological possibilities have multiplied, interest in automation has grown. Seeking to accelerate the development and introduction of fully automated vehicles and highways, the United States Congress authorized more than $650 million over six years for intelligent transport systems (ITS) and demonstration projects in the 1991 Intermodal Surface Transportation Efficiency Act (ISTEA). Congress legislated in ISTEA that "the Secretary of Transportation shall develop an automated highway and vehicle prototype from which future fully automated intelligent vehicle-highway systems can be developed. Such development shall include research in human factors to ensure the success of the man-machine relationship. The goal of this program is to have the first fully automated highway roadway or an automated test track in operation by 1997. This system shall accommodate installation of equipment in new and existing motor vehicles."

Full automation commonly defined as requiring no control or very limited control by the driver; such automation would be accomplished through a combination of sensor, computer, and communications systems in vehicles and along the roadway. Fully automated driving would, in theory, allow closer vehicle spacing and higher speeds, which could enhance traffic capacity in places where additional road building is physically impossible, politically unacceptable, or prohibitively expensive. Automated controls also might enhance road safety by reducing the opportunity for driver error, which causes a large share of motor vehicle crashes. Other potential benefits include improved air quality (as a result of more-efficient traffic flows), increased fuel economy, and spin-off technologies generated during research and development related to automated highway systems.

Automated Waste Management

Automated waste collection trucks prevent the need for as many workers as well as easing the level of labor required to provide the service.

Home Automation

Home automation (also called domotics) designates an emerging practice of increased automation of household appliances and features in residential dwellings, particularly through electronic means that allow for things impracticable, overly expensive or simply not possible in recent past decades.

Industrial Automation

Industrial automation deals primarily with the automation of manufacturing, quality control and material handling processes. General purpose controllers for industrial processes include Programmable logic controllers, stand-alone I/O modules, and computers. One trend is increased use of Machine vision to provide automatic inspection and robot guidance functions, another is a continuing increase in the use of robots.

Energy efficiency in industrial processes has become a higher priority. Semiconductor companies like Infineon Technologies are offering 8-bit micro-controller applications for example found in motor controls, general purpose pumps, fans, and ebikes to reduce energy consumption and thus increase efficiency.

Agriculture

Now that we're moving towards automated orange-sorting and autonomous tractors, the next step in automated agriculture is robotic strawberry pickers.

Agent-assisted automation refers to automation used by call center agents to handle customer inquiries. There are two basic types: desktop automation and automated voice solutions. Desktop automation refers to software programming that makes it easier for the call center agent to work across multiple desktop tools. The automation would take the information entered into one tool and populate it across the others so it did not have to be entered more than once, for example. Automated voice solutions allow the agents to remain on the line while disclosures and other important information is provided to customers in the form of pre-recorded audio files. Specialized applications of these automated voice solutions enable the agents to process credit cards without ever seeing or hearing the credit card numbers or CVV codes

The key benefit of agent-assisted automation is compliance and error-proofing. Agents are sometimes not fully trained or they forget or ignore key steps in the process. The use of automation ensures that what is supposed to happen on the call actually does, every time.

Relationship to Unemployment

Research by the Oxford Martin School showed that employees engaged in "tasks following well-defined procedures that can easily be performed by sophisticated algorithms" are at risk of displacement. The study, published in 2013, shows that automation can affect both skilled and unskilled work and both high and low-paying occupations; however, low-paid physical occupations are most at risk. However, according to a study published in McKinsey Quarterly in 2015 the impact of computerization in most cases is not replacement of employees but automation of portions of the tasks they perform.

Based on a formula by Gilles Saint-Paul, an economist at Toulouse 1 University, the demand for unskilled human capital declines at a slower rate than the demand for skilled human capital increases. In the long run and for society as a whole it has led to cheaper products, lower average work hours, and new industries forming (I.e, robotics industries, computer industries, design industries). These new industries provide many high salary skill based jobs to the economy.

ACTUATOR

An **actuator** is a type of motor that is responsible for moving or controlling a mechanism or system.

It is operated by a source of energy, typically electric current, hydraulic fluid pressure, or pneumatic pressure, and converts that energy into motion. An actuator is the mechanism by which a control system acts upon an environment. The control system can be simple (a fixed mechanical or electronic system), software-based (*e.g.* a printer driver, robot control system), a human, or any other input.

History

The history of the pneumatic actuation system and the hydraulic actuation system dates to around the time of World War II (1938). It was first created by Xhiter Anckeleman (pronounced 'Ziter') who used his knowledge of engines and brake systems to come up with a new solution to ensure that the brakes on a car exert the maximum force, with the least possible wear and tear.

Hydraulic

A hydraulic actuator consists of cylinder or fluid motor that uses hydraulic power to facilitate mechanical operation. The mechanical motion gives an output in terms of linear, rotary or oscillatory motion. Because liquids are nearly impossible to compress, a hydraulic actuator can exert considerable force. The drawback of this approach is its limited acceleration.

The hydraulic cylinder consists of a hollow cylindrical tube along which a piston can slide. The term *single acting* is used when the fluid pressure is applied to just one side of the piston. The piston can move in only one direction, a spring being frequently used to give the piston a return stroke. The term *double acting* is used when pressure is applied on each side of the piston; any difference in pressure between the two side of the piston moves the piston to one side or the other.

Pneumatic

A pneumatic actuator converts energy formed by vacuum or compressed air at high pressure into either linear or rotary motion. Pneumatic energy is desirable for main engine controls because it can quickly respond in starting and stopping as the power source does not need to be stored in reserve for operation.

Fig. : Pneumatic rack and pinion actuators for valve controls of water pipes.

Pneumatic actuators enable large forces to be produced from relatively small pressure changes. These forces are often used with valves to move diaphragms to affect the flow of liquid through the valve.

Electric

An electric actuator is powered by a motor that converts electrical energy into mechanical torque. The electrical energy is used to actuate equipment such as multi-turn valves. It is one of the cleanest and most readily available forms of actuator because it does not involve oil.

Thermal or Magnetic (Shape Memory Alloys)

Actuators which can be actuated by applying thermal or magnetic energy have been used in commercial applications. They tend to be compact, lightweight, economical and with high power density. These actuators use shape memory materials (SMMs), such as shape memory alloys (SMAs) or magnetic shape-memory alloys (MSMAs). Some popular manufacturers of these devices are Finnish Modti Inc. and American Dynalloy.

Mechanical

A mechanical actuator functions by converting rotary motion into linear motion to execute movement. It involves gears, rails, pulleys, chains and other devices to operate. An example is a rack and pinion.

Examples and Applications

In engineering, actuators are frequently used as mechanisms to introduce motion, or to clamp an object so as to prevent motion. In electronic engineering, actuators are a subdivision of transducers. They are devices which transform an input signal (mainly an electrical signal) into motion.

Examples of Actuators :

- Comb drive
- Digital micromirror device
- Electric motor
- Electroactive polymer
- Hydraulic cylinder
- Piezoelectric actuator
- Pneumatic actuator
- Servomechanism
- Thermal bimorph
- Screw jack

Circular to Linear Conversion

Motors are mostly used when circular motions are needed, but can also be used for linear applications by transforming circular to linear motion with a lead screw or similar mechanism. On the other hand, some actuators are intrinsically linear, such as piezoelectric actuators. Conversion between circular and linear motion is commonly made via a few simple types of mechanism including:

- **Screw:** Screw jack, ball screw and roller screw actuators all operate on the principle of the simple machine known as the screw. By rotating the actuator's nut, the screw shaft moves in a line. By moving the screw shaft, the nut rotates.

- **Wheel and axle:** Hoist, winch, rack and pinion, chain drive, belt drive, rigid chain and rigid belt actuators operate on the principle of the wheel and axle. By rotating a wheel/axle (*e.g.* drum, gear, pulley or shaft) a linear member (*e.g.* cable, rack, chain or belt) moves. By moving the linear member, the wheel/axle rotates.

Virtual Instrumentation

In virtual instrumentation, actuators and sensors are the hardware complements of virtual instruments.

Performance Metrics

Performance metrics for actuators include speed, acceleration, and force (alternatively, angular speed, angular acceleration, and torque), as well as energy efficiency and considerations such as mass, volume, operating conditions, and durability, among others.

Force

When considering force in actuators for applications, two main metrics should be considered. These two are static and dynamic loads. Static load is the force capability of the actuator while not in motion. Conversely, the dynamic load of the actuator is the force capability while in motion. The two aspects rarely have the same weight capability and must be considered separately.

Speed

Speed should be considered primarily at a no-load pace, since the speed will invariably decrease as the load amount increases. The rate the speed will decrease will directly correlate with the amount of force and the initial speed.

Operating Conditions

Actuators are commonly rated using the standard IP Code rating system. Those that are rated for dangerous environments will have a higher IP rating than those for personal or common industrial use.

Durability

This will be determined by each individual manufacturer, depending on usage and quality.

CONTROL ENGINEERING

Control engineering or **control systems engineering** is the engineering discipline that applies control theory to design systems with desired behaviors. The practice uses sensors to measure the output performance of the device being controlled and those measurements can be used to give feedback to the input actuators that can make corrections toward desired performance. When a device is designed to perform without the need of human inputs for correction it is called automatic control (such as cruise control for regulating the speed of a car). Multi-disciplinary in nature, control systems engineering activities focus on implementation of control systems mainly derived by mathematical modeling of systems of a diverse range.

Modern day control engineering is a relatively new field of study that gained significant attention during the 20th century with the advancement of technology. It can be broadly defined or classified as practical application of control theory. Control engineering has an essential role in a wide range of control systems, from simple household washing machines to high-performance F-16 fighter aircraft. It seeks to understand physical systems, using mathematical modeling, in terms of inputs, outputs and various components with different behaviors, use control systems design tools to develop controllers for those systems and implement controllers in physical systems employing available technology. A system can be mechanical, electrical, fluid, chemical, financial and even biological, and the mathematical modeling, analysis and controller design uses control theory in one or many of the time, frequency and complex-s domains, depending on the nature of the design problem.

Fig. : Control systems play a critical role in space flight.

Automatic control systems were first developed over two thousand years ago. The first feedback control device on record is thought to be the ancient Ktesibios's water clock in Alexandria, Egypt around the third century B.C. It kept time by regulating the water level in a vessel and, therefore, the water flow from that vessel. This certainly was a successful device as water clocks of similar design were still being made in Baghdad when the Mongols captured the city in 1258 A.D.

A variety of automatic devices have been used over the centuries to accomplish useful tasks or simply to just entertain. The latter includes the automata, popular in Europe in the 17th and 18th centuries, featuring dancing figures that would repeat the same task over and over again; these automata are examples of open-loop control. Milestones among feedback, or "closed-loop" automatic control devices, include the temperature regulator of a furnace attributed to Drebbel, circa 1620, and the centrifugal flyball governor used for regulating the speed of steam engines by James Watt in 1788.

In his 1868 paper "On Governors", James Clerk Maxwell was able to explain instabilities exhibited by the flyball governor using differential equations to describe the control system. This demonstrated the importance and usefulness of mathematical models and methods in understanding complex phenomena, and signaled the beginning of mathematical control and systems theory. Elements of control theory had appeared earlier but not as dramatically and convincingly as in Maxwell's analysis.

Control theory made significant strides in the next 100 years. New mathematical techniques made it possible to control, more accurately, significantly more complex dynamical systems than the original flyball governor. These techniques include developments in optimal control in the 1950s and 1960s, followed by progress in stochastic, robust, adaptive and optimal control methods in the 1970s and 1980s. Applications of control methodology have helped make possible space travel and communication satellites, safer and more efficient aircraft, cleaner auto engines, cleaner and more efficient chemical processes.

Before it emerged as a unique discipline, control engineering was practiced as a part of mechanical engineering and control theory was studied as a part of electrical engineering since electrical circuits can often be easily described using control theory techniques. In the very first control relationships, a current output was represented with a voltage control input. However, not having proper technology to implement electrical control systems, designers left with the option of less efficient and slow responding mechanical systems. A very effective mechanical controller that is still widely used in some hydro plants is the governor. Later on, previous to modern power electronics, process control systems for industrial applications were devised by mechanical engineers using pneumatic and hydraulic control devices, many of which are still in use today.

Control Theory

There are two major divisions in control theory, namely, classical and modern, which have direct implications over the control engineering applications.

The scope of classical control theory is limited to single-input and single-output (SISO) system design, except when analyzing for disturbance rejection using a second input. The system analysis is carried out in the time domain using differential equations, in the complex-s domain with the Laplace transform, or in the frequency domain by transforming from the complex-s domain. Many systems may be assumed to have a second order and single variable system response in the time domain.

A controller designed using classical theory often requires on-site tuning due to incorrect design approximations. Yet, due to the easier physical implementation of classical controller designs as compared to systems designed using modern control theory, these controllers are preferred in most industrial applications. The most common controllers designed using classical control theory are PID controllers. A less common implementation may include either or both a Lead or Lag filter.

The ultimate end goal is to meet a requirements set typically provided in the time-domain called the Step response, or at times in the frequency domain called the Open-Loop response. The Step response characteristics applied in a specification are typically percent overshoot, settling time, *etc*. The Open-Loop response characteristics applied in a specification are typically Gain and Phase margin and bandwidth. These characteristics may be evaluated through simulation including a dynamic model of the system under control coupled with the compensation model.

In contrast, modern control theory is carried out in the state space, and can deal with multi-input and multi-output (MIMO) systems. This overcomes the limitations of classical control theory in more sophisticated design problems, such as fighter aircraft control, with the limitation that no frequency domain analysis is possible. In modern design, a system is represented to the greatest advantage as a set of decoupled first order differential equations defined using state variables.

Nonlinear, multivariable, adaptive and robust control theories come under this division. Matrix methods are significantly limited for MIMO systems where linear independence cannot be assured in the relationship between inputs and outputs. Being fairly new, modern control theory has many areas yet to be explored. Scholars like Rudolf E. Kalman and Aleksandr Lyapunov are well-known among the people who have shaped modern control theory.

Control Systems

Control engineering is the engineering discipline that focuses on the modeling of a diverse range of dynamic systems (*e.g.* mechanical systems) and the design of controllers that will cause these systems to behave in the desired manner. Although such controllers need not be electrical many are and hence control engineering is often viewed as a subfield of electrical engineering. However, the falling price of microprocessors is making the actual implementation of a control system essentially trivial. As a result, focus is shifting back to the mechanical and process engineering discipline, as intimate knowledge of the physical system being controlled is often desired.

Electrical circuits, digital signal processors and microcontrollers can all be used to implement control systems. Control engineering has a wide range of applications from the flight and propulsion systems of commercial airliners to the cruise control present in many modern automobiles.

In most of the cases, control engineers utilize feedback when designing control systems. This is often accomplished using a PID controller system. For example, in an automobile with cruise control the vehicle's speed is continuously monitored and fed back to the system, which adjusts the motor's torque accordingly. Where there is regular feedback, control theory can be used to determine how the system responds to such feedback. In practically all such systems stability is important and control theory can help ensure stability is achieved.

Although feedback is an important aspect of control engineering, control engineers may also work on the control of systems without feedback. This is known as open loop control. A classic example of open loop control is a washing machine that runs through a pre-determined cycle without the use of sensors.

Control Engineering Education

At many universities, control engineering courses are taught in electrical and electronic engineering, mechatronics engineering, mechanical engineering, and aerospace engineering. In others, control engineering is connected to computer science, as most control techniques today are implemented through computers, often as embedded systems (as in the automotive field). The field of control within chemical engineering is often known as process control. It deals primarily with the control of variables in a chemical process in a plant.

It is taught as part of the undergraduate curriculum of any chemical engineering program and employs many of the same principles in control engineering. Other engineering disciplines also overlap with control engineering as it can be applied to any system for which a suitable model can be derived. However, specialised control engineering departments do exist, for example, the Department of Automatic Control and Systems Engineering at the University of Sheffield and the Department of Systems Engineering at the United States Naval Academy.

Control engineering has diversified applications that include science, finance management, and even human behavior. Students of control engineering may start with a linear control system course dealing with the time and complex-s domain, which requires a thorough background in elementary mathematics and Laplace transform, called classical control theory. In linear control, the student does frequency and time domain analysis. Digital control and nonlinear control courses require Z transformation and algebra respectively, and could be said to complete a basic control education.

Recent Advancement

Originally, control engineering was all about continuous systems. Development of computer control tools posed a requirement of discrete control system engineering because the communications between the computer-based digital controller and the physical system are governed by a computer clock. The equivalent to Laplace transform in the discrete domain is the Z-transform. Today, many of the control systems are computer controlled and they consist of both digital and analog components.

Therefore, at the design stage either digital components are mapped into the continuous domain and the design is carried out in the continuous domain, or analog components are mapped into discrete domain and design is carried out there. The first of these two methods is more commonly encountered in practice because many industrial systems have many continuous systems components, including mechanical, fluid, biological and analog electrical components, with a few digital controllers.

Similarly, the design technique has progressed from paper-and-ruler based manual design to computer-aided design and now to computer-automated design or CAutoD which has been made possible by evolutionary computation. CAutoD can be applied not just to tuning a predefined control scheme, but also to controller structure optimisation, system identification and invention of novel control systems, based purely upon a performance requirement, independent of any specific control scheme.

Resilient Control Systems extends the traditional focus on addressing only plant disturbances to frameworks, architectures and methods that address multiple types of unexpected disturbance. In particular, adapting and transforming behaviors of the control system in response to malicious actors, abnormal failure modes, undesirable human action, *etc.* Development of resilience technologies require the involvement of multidisciplinary teams to holistically address the performance challenges.

DISTRIBUTED CONTROL SYSTEM

A **distributed control system (DCS)** is a control system for a process or plant, wherein control elements are distributed throughout the system. This is in contrast to non-distributed systems, which use a single controller at a central location. In a DCS, a hierarchy of controllers is connected by communications networks for command and monitoring.

Example scenarios where a DCS might be used include:

- Chemical plants
- Petrochemical (oil) and refineries
- Pulp and Paper Mills
- Boiler controls and power plant systems
- Nuclear power plants
- Environmental control systems
- Water management systems

- Metallurgical process plants
- Pharmaceutical manufacturing
- Sugar refining plants
- Dry cargo and bulk oil carrier ships
- Formation control of multi-agent systems

Elements

A DCS typically uses custom designed processors as controllers and uses both proprietary interconnections and standard communications protocol for communication. Input and output modules form component parts of the DCS. The processor receives information from input modules and sends information to output modules. The input modules receive information from input instruments in the process (or field) and the output modules transmit instructions to the output instruments in the field.

The inputs and outputs can be either analog signal which are continuously changing or discrete signals which are 2 state either on or off . Computer buses or electrical buses connect the processor and modules through multiplexer or demultiplexers. Buses also connect the distributed controllers with the central controller and finally to the Human–machine interface (HMI) or control consoles.

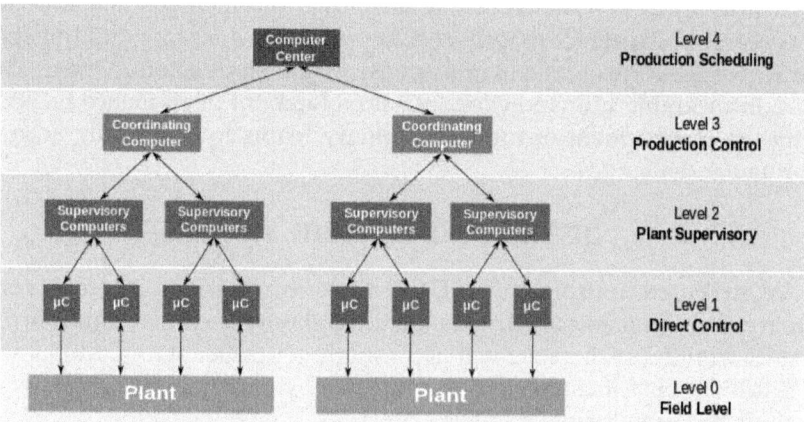

Fig. : Functional levels of a typical Distributed Control System.

The elements of a DCS may connect directly to physical equipment such as switches, pumps and valves and to Human Machine Interface (HMI) via SCADA. The differences between a DCS and SCADA is often subtle, especially with advances in technology allowing the functionality of each to overlap.

Applications

Distributed control systems (DCSs) are dedicated systems used to control manufacturing processes that are continuous or batch-oriented, such as oil refining, petrochemicals, central station power generation, fertilizers, pharmaceuticals, food and beverage manufacturing, cement production, steelmaking, and papermaking. DCSs are connected to sensors and actuators and use setpoint control to control the flow of material through the plant. The most common example is a setpoint control loop consisting of a pressure sensor, controller, and control valve. Pressure or flow measurements are transmitted to the controller, usually through the aid of a signal conditioning input/output (I/O) device.

When the measured variable reaches a certain point, the controller instructs a valve or actuation device to open or close until the fluidic flow process reaches the desired setpoint. Large oil refineries have many thousands of I/O points and employ very large DCSs. Processes are not limited to fluidic flow through pipes, however, and can also include things like paper machines and their associated quality controls, variable speed drives and motor control centers, cement kilns, mining operations, ore processing facilities, and many others.

A typical DCS consists of functionally and/or geographically distributed digital controllers capable of executing from 1 to 256 or more regulatory control loops in one control box. The input/output devices (I/O) can be integral with the controller or located remotely via a field network. Today's controllers have extensive computational capabilities and, in addition to proportional, integral, and derivative (PID) control, can generally perform logic and sequential control. Modern DCSs also support neural networks and fuzzy application. Recent research focuses on the synthesis of optimal distributed controllers, which optimizes a certain H-infinity or H-2 criterion.

DCSs are usually designed with redundant processors to enhance the reliability of the control system. Most systems come with displays and configuration software that enable the end-user to configure the control system without the need for performing low-level programming, allowing the user also to better focus on the application rather than the equipment. However, considerable system knowledge and skill is required to properly deploy the hardware, software, and applications. Many plants have dedicated personnel who focus on these tasks, augmented by vendor support that may include maintenance support contracts.

DCSs may employ one or more workstations and can be configured at the workstation or by an off-line personal computer. Local communication is handled by a control network with transmission over twisted -pair, coaxial, or fiber-optic cable. A server and/or applications processor may be included in the system for extra computational, data collection, and reporting capability.

Early minicomputers were used in the control of industrial processes since the beginning of the 1960s. The IBM 1800, for example, was an early computer that had input/output hardware to gather process signals in a plant for conversion from field contact levels (for digital points) and analog signals to the digital domain.

The first industrial control computer system was built 1959 at the Texaco Port Arthur, Texas, refinery with an RW-300 of the Ramo-Wooldridge Company

In 1975, both Honeywell and Japanese electrical engineering firm Yokogawa introduced their own independently produced DCS's with Yokogawa introducing and successfully installing before Honeywell, with the TDC 2000 and CENTUM systems, respectively. US-based Bristol also introduced their UCS 3000 universal controller in 1975. In 1978 Metso(known as Valmet in 1978) introduced their own DCS system called Damatic (latest generation named Metso DNA). In 1980, Bailey (now part of ABB) introduced the NETWORK 90 system, Fisher Controls (now part of Emerson Electric) introduced the PROVoX system, Fischer & Porter Company (now also part of ABB) introduced DCI-4000 (DCI stands for Distributed Control Instrumentation).

The DCS largely came about due to the increased availability of microcomputers and the proliferation of microprocessors in the world of process control. Computers had already been applied to process automation for some time in the form of both direct digital control (DDC) and set point control. In the early 1970s Taylor Instrument Company, (now part of ABB) developed the 1010 system, Foxboro the FOX1 system, Fisher Controls the DC system and Bailey Controls the 1055 systems. All of these were DDC applications implemented within minicomputers (DEC PDP-11, Varian Data Machines, MODCOMP *etc.*) and connected to proprietary Input/Output hardware. Sophisticated (for the time) continuous as well as batch control was implemented in this way. A more conservative approach was set point control, where process computers supervised clusters of analog process controllers. A CRT-based workstation provided visibility into the process using text and crude character graphics. Availability of a fully functional graphical user interface was a way away.

Central to the DCS model was the inclusion of control function blocks. Function blocks evolved from early, more primitive DDC concepts of "Table Driven" software. One of the first embodiments of object-oriented software, function blocks were self-contained "blocks" of code that emulated analog hardware control components and performed tasks that were essential to process control, such as execution of PID algorithms. Function blocks continue to endure as the predominant method of control for DCS suppliers, and are supported by key technologies such as Foundation Fieldbus today.

Midac Systems, of Sydney, Australia, developed an objected-oriented distributed direct digital control system in 1982. The central system ran 11 microprocessors sharing tasks and common memory and connected to a serial communication network of distributed controllers each running two Z80s. The system was installed at the University of Melbourne.

Digital communication between distributed controllers, workstations and other computing elements (peer to peer access) was one of the primary advantages of the DCS. Attention was duly focused on the networks, which provided the all-important lines of communication that, for process applications, had to incorporate specific functions such as determinism and redundancy. As a result, many suppliers embraced the IEEE 802.4 networking standard. This decision set the stage for the wave of migrations necessary when information technology moved into process automation and IEEE 802.3 rather than IEEE 802.4 prevailed as the control LAN.

The Network-centric Era of the 1980s

In the 1980s, users began to look at DCSs as more than just basic process control. A very early example of a Direct Digital Control DCS was completed by the Australian business Midac in 1981–82 using R-Tec Australian designed hardware. The system installed at the University of Melbourne used a serial communications network, connecting campus buildings back to a control room "front end". Each remote unit ran two Z80 microprocessors, while the front end ran eleven Z80s in a parallel processing configuration with paged common memory to share tasks and that could run up to 20,000 concurrent control objects.

It was believed that if openness could be achieved and greater amounts of data could be shared throughout the enterprise that even greater things could be achieved. The first attempts to increase the openness of DCSs resulted in the adoption of the predominant operating system of the day: *UNIX*. UNIX and its companion networking technology TCP-IP were developed by the US Department of Defense for openness, which was precisely the issue the process industries were looking to resolve.

As a result, suppliers also began to adopt Ethernet-based networks with their own proprietary protocol layers. The full TCP/IP standard was not implemented, but the use of Ethernet made it possible to implement the first instances of object management and global data access technology. The 1980s also witnessed the first PLCs integrated into the DCS infrastructure. Plant-wide historians also emerged to capitalize on the extended reach of automation systems. The first DCS supplier to adopt UNIX and Ethernet networking technologies was Foxboro, who introduced the I/A Series system in 1987.

The Application-centric Era of the 1990s

The drive toward openness in the 1980s gained momentum through the 1990s with the increased adoption of commercial off-the-shelf (COTS) components and IT standards. Probably the biggest transition undertaken during this time was the move from the UNIX operating system to the Windows environment. While the realm of the real time operating system (RTOS) for control applications remains dominated by real time commercial variants of UNIX or proprietary operating systems, everything above real-time control has made the transition to Windows.

The introduction of Microsoft at the desktop and server layers resulted in the development of technologies such as OLE for process control (OPC), which is now a de facto industry connectivity standard. Internet technology also began to make its mark in automation and the DCS world, with most DCS HMI supporting Internet connectivity. The 1990s were also known for the "Fieldbus Wars", where rival organizations competed to define what would become the IEC fieldbus standard for digital communication with field instrumentation instead of 4–20 milliamp analog communications. The first fieldbus installations occurred in the 1990s.

Towards the end of the decade, the technology began to develop significant momentum, with the market consolidated around Ethernet I/P, Foundation Fieldbus and Profibus PA for process automation applications. Some suppliers built new systems from the ground up to maximize functionality with fieldbus, such as Rockwell PlantPAX System, Honeywell with Experion & Plantscape SCADA systems, ABB with System 800xA, Emerson Process Management with the Emerson Process Management DeltaV control system, Siemens with the SPPA-T3000 or Simatic PCS 7, Forbes Marshall with the Microcon+ control system and Azbil Corporation with the Harmonas-DEO system. Fieldbus technics have been used to integrate machine, drives, quality and condition monitoring applications to one DCS with Metso DNA system.

The impact of COTS, however, was most pronounced at the hardware layer. For years, the primary business of DCS suppliers had been the supply of large amounts of hardware, particularly I/O and controllers. The initial proliferation of DCSs required the installation of prodigious amounts of this hardware, most of it manufactured from the bottom up by DCS suppliers. Standard computer components from manufacturers such as Intel and Motorola, however, made it cost prohibitive for DCS suppliers to continue making their own components, workstations, and networking hardware.

As the suppliers made the transition to COTS components, they also discovered that the hardware market was shrinking fast. COTS not only resulted in lower manufacturing costs for the supplier, but also steadily decreasing prices for the end users, who were also becoming increasingly vocal over what they perceived to be unduly high hardware costs. Some suppliers that were previously stronger in the PLC business, such as Rockwell Automation and Siemens, were able to leverage their expertise in manufacturing control hardware to enter the DCS marketplace with cost effective offerings, while the stability/scalability/reliability and functionality of these emerging systems are still improving.

The traditional DCS suppliers introduced new generation DCS System based on the latest Communication and IEC Standards, which resulting in a trend of combining the traditional concepts/functionalities for PLC and DCS into a one for all solution — named "Process Automation System". The gaps among the various systems remain at the areas such as: the database integrity, pre-engineering functionality, system maturity, communication transparency and reliability. While it is expected the cost ratio is relatively the same (the more powerful the systems are, the more expensive they will be), the reality of the automation business is often operating strategically case by case. The current next evolution step is called Collaborative Process Automation Systems.

To compound the issue, suppliers were also realizing that the hardware market was becoming saturated. The life cycle of hardware components such as I/O and wiring is also typically in the range of 15 to over 20 years, making for a challenging replacement market. Many of the older systems that were installed in the 1970s and 1980s are still in use today, and there is a considerable installed base of systems in the market that are approaching the end of their useful life. Developed industrial economies in North America, Europe, and Japan already had many thousands of DCSs installed, and with few if any new plants being built, the market for new hardware was shifting rapidly to smaller, albeit faster growing regions such as China, Latin America, and Eastern Europe.

Because of the shrinking hardware business, suppliers began to make the challenging transition from a hardware-based business model to one based on software and value-added services. It is a transition that is still being made today. The applications portfolio offered by suppliers expanded considerably in the '90s to include areas such as production management, model-based control, real-time optimization, plant asset management (PAM), Real-time performance management (RPM) tools, alarm management, and many others. To obtain the true value from these applications, however, often requires a considerable service content, which the suppliers also provide.

Modern Systems (2010 onwards)

The latest developments in DCS include the following new technologies:

1. Wireless systems and protocols
2. Remote transmission, logging and data historian
3. Mobile interfaces and controls
4. Embedded web-servers

Increasingly, and ironically, DCS are becoming centralised at plant level, with the ability to log in to remote equipment. This enables the provision of a superior human-machine interface (HMI) especially from the point of view of remote access and portability.

As wireless protocols are developed and refined, DCS increasingly includes wireless communication. DCS controllers are now often equipped with embedded servers and provide on-the-go web access.

Many vendors provide the option of a mobile HMI, ready for both Android and iOS. With these interfaces, the threat of security breaches and possible damage to plant and process are now very real.

Chapter 2

FLOW CONTROL VALVE

A flow control valve regulates the flow or pressure of a fluid. Control valves normally respond to signals generated by independent devices such as flow meters or temperature gauges.

Control valves are normally fitted with actuators and positioners. Pneumatically-actuated globe valves and Diaphragm Valves are widely used for control purposes in many industries, although quarter-turn types such as (modified) ball, gate and butterfly valves are also used.

Control valves can also work with hydraulic actuators (also known as hydraulic pilots). These types of valves are also known as Automatic Control Valves. The hydraulic actuators will respond to changes of pressure or flow and will open/close the valve. Automatic Control Valves do not require an external power source, meaning that the fluid pressure is enough to open and close the valve.

Automatic control valves include: pressure reducing valves, flow control valves, back-pressure sustaining valves, altitude valves, and relief valves. An altitude valve controls the level of a tank. The altitude valve will remain open while the tank is not full and it will close when the tanks reaches its maximum level. The opening and closing of the valve requires no external power source (electric, pneumatic, or man power), it is done automatically, hence its name.

Process plants consist of hundreds, or even thousands, of control loops all networked together to produce a product to be offered for sale. Each of these control loops is designed to keep some important process variable such as pressure, flow, level, temperature, *etc.* within a required operating range to ensure the quality of the end product. Each of these loops receives and internally creates disturbances that detrimentally affect the process variable, and interaction from other loops in the network provides disturbances that influence the process variable.

To reduce the effect of these load disturbances, sensors and transmitters collect information about the process variable and its relationship to some desired set point. A controller will then process this information and decides what must be

Fig. : Globe control valve with the pneumatic actuator and smart positioner

done to get the process variable back to where it should be after a load disturbance occurs. When all the measuring, comparing, and calculating are done, some type of final control element must implement the strategy selected by the controller. The most common final control element in the process control industries is the control valve. The control valve manipulates a flowing fluid, such as gas, steam, water, or chemical compounds, to compensate for the load disturbance and keep the regulated process variable as close as possible to the desired set point.

Temperature Control

Temperature control is a process in which change of temperature of a space (and objects collectively there within) is measured or otherwise detected, and the passage of heat energy into or out of the space is adjusted to achieve a desired average temperature.

Fig. : Temperature measuring and controlling module for microcontroller experiment.

Control Loops

A home thermostat is an example of a closed control loop: It constantly assesses the current room temperature and controls a heater and/or air conditioner to increase or decrease the temperature according to user-defined setting(s). A simple (low-cost, cheap) thermostat merely switches the heater or air conditioner either on or off, and temporary overshoot and undershoot of the desired average temperature must be expected.

A more expensive thermostat varies the amount of heat or cooling provided by the heater or cooler, depending on the difference between the required temperature (the "setpoint") and the actual temperature. This minimizes over/undershoot. This method is called Proportional control. Further enhancements using the accumulated error signal (Integral) and the rate at which the error is changing (Derivative) are used to form more complex PID Controllers which is the form usually seen in industry.

Energy Balance

An object's or space's temperature increases when heat energy moves into it, increasing the average kinetic energy of its atoms, *e.g.*, of things and air in a room. Heat energy leaving an object or space lowers its temperature. Heat flows from one place to another (always from a higher temperature to a lower one) by one or more of three processes: conduction, convection and radiation. In conduction, energy is passed from one atom to another by direct contact.

In convection, heat energy moves by conduction into some movable fluid (such as air or water) and the fluid moves from one place to another, carrying the heat with it. At some point the heat energy in the fluid is usually transferred to some other object by means conduction again. The movement of the fluid can be driven by negative-buoyancy, as when cooler (and therefore denser) air drops and thus upwardly displaces warmer (less-dense) air (natural convection), or by fans or pumps (forced convection).

In radiation, the heated atoms make electromagnetic emissions absorbed by remote other atoms, whether nearby or at astronomical distance. For example, the Sun radiates heat as both invisible and visible electromagnetic energy. What we know as "light" is but a narrow region of the electromagnetic spectrum.

If, in a place or thing, more energy is received than is lost, its temperature increases. If the amount of energy coming in and going out are exactly the same, the temperature stays constant — there is thermal balance, or thermal equilibrium.

Servomechanism

A **servomechanism**, sometimes shortened to **servo**, is an automatic device that uses error-sensing negative feedback to correct the performance of a mechanism and is defined by its function. It usually includes a built-in encoder. A servomechanism is sometimes called a **heterostat** since it controls a system's behavior by means of heterostasis.

Fig. : Industrial servomotor.

The grey/green cylinder is the brush-type DC motor. The black section at the bottom contains the planetary reduction gear, and the black object on top of the motor is the optical rotary encoder for position feedback. This is the steering actuator of a large robot vehicle.

The term correctly applies only to systems where the feedback or error-correction signals help control mechanical position, speed or other parameters. For example, an automotive power window control is not a servomechanism, as there is no automatic feedback that controls position — the operator does this by observation. By contrast a car's cruise control uses closed loop feedback, which classifies it as a servomechanism.

Uses

Position Control

A common type of servo provides *position control*. Servos are commonly electrical or partially electronic in nature, using an electric motor as the primary means of creating mechanical force. Other types of servos use hydraulics, pneumatics, or magnetic principles. Servos operate on the principle of negative feedback, where the control input is compared to the actual position of the mechanical system as measured by some sort of transducer at the output. Any difference between the actual and wanted values (an "error signal") is amplified (and converted) and used to drive the system in the direction necessary to reduce or eliminate the error. This procedure is one widely used application of control theory.

Speed Control

Speed control via a governor is another type of servomechanism. The steam engine uses mechanical governors; another early application was to govern the speed of water wheels. Prior to World War II the constant speed propeller was developed to control engine speed for maneuvering aircraft. Fuel controls for gas turbine engines employ either hydromechanical or electronic governing.

Other

Positioning servomechanisms were first used in military fire-control and marine navigation equipment. Today servomechanisms are used in automatic machine tools, satellite-tracking antennas, remote control airplanes, automatic navigation systems on boats and planes, and antiaircraft-gun control systems. Other examples are fly-by-wire systems in aircraft which use servos to actuate the aircraft's control surfaces, and radio-controlled models which use RC servos for the same purpose. Many autofocus cameras also use a servomechanism to accurately move the lens, and thus adjust the focus. A modern hard disk drive has a magnetic servo system with sub-micrometre positioning accuracy. In industrial machines, servos are used to perform complex motion, in many applications.

Rotary or Linear

Typical servos give a rotary (angular) output. Linear types are common as well, using a leadscrew or a linear motor to give linear motion.

Servomotor

A *servomotor* is a specific type of motor that is combined with a rotary encoder or a potentiometer to form a servomechanism. This assembly may in turn form part of another servomechanism. A potentiometer provides a simple analog signal to indicate position, while an encoder provides position and usually speed feedback, which by the use of a PID controller allow more precise control of position and thus faster achievement of a stable position (for a given motor power). Potentiometers are subject to drift when the temperature changes whereas encoders are more stable and accurate.

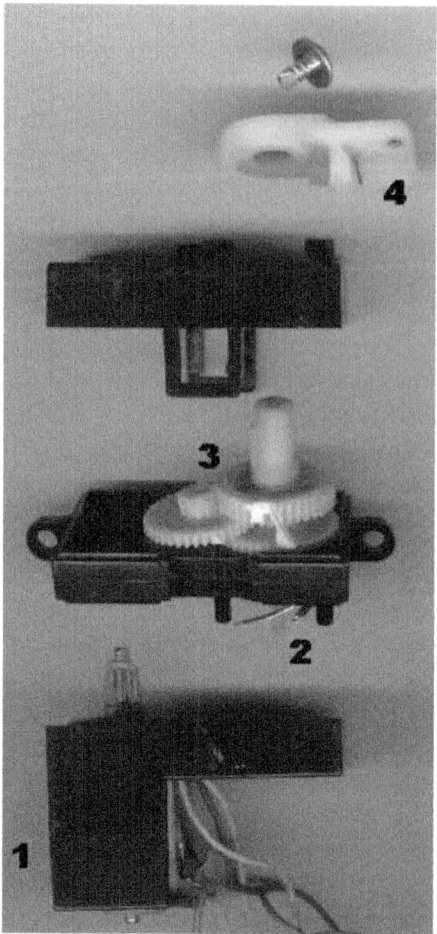

Fig. : Small R/C servo mechanism.
1. electric motor
2. position feedback potentiometer
3. reduction gear
4. actuator arm

Servomotors are used for both high-end and low-end applications. On the high end are precision industrial components that use a rotary encoder. On the low end are inexpensive radio control servos (RC servos) used in radio-controlled models which use a free-running motor and a simple potentiometer position sensor with an embedded controller. The term *servomotor* generally refers to a high-end industrial component while the term *servo* is most often used to describe the inexpensive devices that employ a potentiometer. Stepper motors are not considered to be servomotors, although they too are used to construct larger servomechanisms. Stepper motors have inherent angular positioning, owing to their construction, and this is generally used in an open-loop manner without feedback. They are generally used for medium-precision applications.

RC servos are used to provide actuation for various mechanical systems such as the steering of a car, the control surfaces on a plane, or the rudder of a boat. Due to their affordability, reliability, and simplicity of control by microprocessors, they are often used in small-scale robotics applications. A standard RC receiver (or a microcontroller) sends pulse-width modulation (PWM) signals to the servo. The electronics inside the servo translate the width of the pulse into a position. When the servo is commanded to rotate, the motor is powered until the potentiometer reaches the value corresponding to the commanded position.

REGULATOR (AUTOMATIC CONTROL)

In automatic control, a **regulator** is a device which has the function of maintaining a designated characteristic. It performs the activity of managing or maintaining a range of values in a machine. The measurable property of a device is managed closely by specified conditions or an advance set value; or it can be a variable according to a predetermined arrangement scheme. It can be used generally to connote any set of various controls or devices for regulating or controlling items or objects.

Examples are a voltage regulator (which can be a transformer whose voltage ratio of transformation can be adjusted, or an electronic circuit that produces a defined voltage), a pressure regulator, such as a diving regulator, which maintains its output at a fixed pressure lower than its input, and a fuel regulator (which controls the supply of fuel).

Regulators can be designed to control anything from gases or fluids, to light or electricity. Speed can be regulated by electronic, mechanical, or electro-mechanical means. Such instances include;

- Electronic regulators as used in modern railway sets where the voltage is raised or lowered to control the speed of the engine
- Mechanical systems such as valves as used in fluid control systems. Purely mechanical pre-automotive systems included such designs as the Watt centrifugal governor whereas modern systems may have electronic fluid speed sensing components directing solenoids to set the valve to the desired rate.
- Complex electro-mechanical speed control systems used to maintain speeds in modern cars (cruise control) - often including hydraulic components,
- An aircraft engine's constant speed unit changes the propellor pitch to maintain engine speed.

GOVERNOR (DEVICE)

A **governor**, or **speed limiter**, is a device used to measure and regulate the speed of a machine, such as an engine. A classic example is the centrifugal governor, also known as the Watt or fly-ball governor, which uses weights mounted on spring-loaded arms to determine how fast a shaft is spinning, and then uses proportional control to regulate the shaft speed.

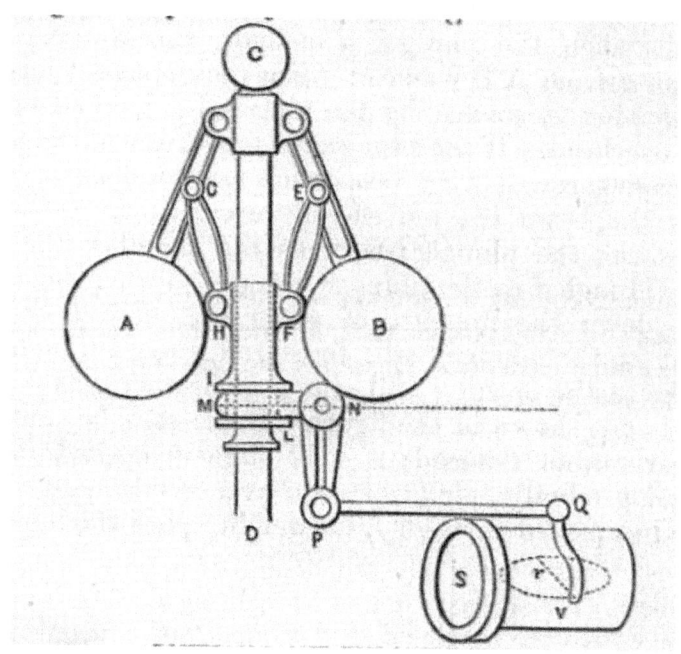

Fig. : Schematic of Engine Governor.

Centrifugal governors were used to regulate the distance and pressure between millstones in windmills since the 17th century. Early steam engines employed a purely reciprocating motion, and were used for pumping water – an application that could tolerate variations in the working speed. It was not until the Scottish engineer James Watt introduced the *rotative* steam engine, for driving factory machinery, that a constant operating speed became necessary.

Between the years 1775 and 1800, Watt, in partnership with industrialist Matthew Boulton, produced some 500 rotative beam engines. At the heart of these engines was Watt's self-designed "conical pendulum" governor: a set of revolving steel balls attached to a vertical spindle by link arms, where the controlling force consists of the weight of the balls. The theoretical basis for the operation of governors was described by James Clerk Maxwell in 1868 in his seminal paper 'On Governors'.

Building on Watt's design was American engineer Willard Gibbs who in 1872 theoretically analyzed Watt's conical pendulum governor from a mathematical energy balance perspective. During his Graduate school years at Yale University, Gibbs observed that the operation of the device in practice was beset with the disadvantages of sluggishness and a tendency to overcorrect for the changes in speed it was supposed to control. Gibbs theorized that, analogous to the equilibrium of the simple Watt governor (which depends on the balancing of two torques: one due to the weight of the "balls" and the other due to their rotation), thermodynamic equilibrium for any work producing thermodynamic system depends on the balance of two entities.

The first is the heat energy supplied to the intermediate substance, and the second is the work energy performed by the intermediate substance. In this case, the intermediate substance is steam. These sorts of theoretical investigations culminated in the 1876 publication of the Gibbs' famous work *On the Equilibrium of Heterogeneous Substances* and in the construction of the Gibbs' governor, shown adjacent. These formulations are ubiquitous today in the natural sciences in the form of the Gibbs' free energy equation, which is used to determine the equilibrium of chemical reactions; also known as *Gibbs equilibrium*.

Speed Limiters

Governors can be used to limit the top speed for vehicles, and for some classes of vehicle such devices are a legal requirement. They can more generally be used to limit the rotational speed of the internal combustion engine or protect the engine from damage due to excessive rotational speed.

Cars

Today, BMW, Audi, Volkswagen and Mercedes-Benz limit their production cars to 250 kilometres per hour (155 mph). Certain Quattro GmbH and AMG cars, and the Mercedes/McLaren SLR is an exception. The BMW Rolls-Royces are limited to 240 kilometres per hour (149 mph). Jaguars, although British, also have a limiter, as do the Swedish Saab and Volvo on cars where it is necessary.

German manufacturers initially started the "gentlemen's agreement", electronically limiting their vehicles to a top speed of 250 kilometres per hour (155 mph), since such high speeds are more likely on the Autobahn. This was done to reduce the political willpower to introduce a legal speed limit.

In European markets, General Motors Europe sometimes choose to discount the agreement, meaning that certain high-powered Opel or Vauxhall cars can exceed the 250 kilometres per hour (155 mph) mark, whereas their Cadillacs do not. Ferrari, Lamborghini, Maserati, Porsche, Aston Martin and Bentley also do not limit their cars, at least not to 250 kilometres per hour (155 mph). The Chrysler 300C SRT8 is limited to 270 km/h. Most Japanese domestic market vehicles are limited to only 180 kilometres per hour (112 mph) or 190 kilometres per hour (118 mph). The top speed is a strong sales argument, though speeds above about 300 kilometres per hour (190 mph) are not likely reachable on public roads.

Many performance cars are limited to a speed of 250 kilometres per hour (155 mph) to limit insurance costs of the vehicle, and reduce the risk of tires failing.

Mopeds

Mopeds in the United Kingdom have had to have a 30 mph (48 km/h) speed limiter since 1977. Most other European countries have similar rules.

Public Services Vehicles

Public service vehicles often have a legislated top speed. Scheduled coach services in the United kingdom (and also bus services) are limited to 65 mph.

Urban public buses often have speed governors which are typically set to between 65 kilometres per hour (40 mph) and 100 kilometres per hour (62 mph).

Trucks (HGVs)

All heavy vehicles in Europe and New Zealand have law/by-law governors that limits their speeds to 90 kilometres per hour (56 mph) or 100 kilometres per hour (62 mph).

Example Uses

Aircraft

Aircraft propellers are another application. The governor senses shaft RPM, and adjusts or controls the angle of the blades to vary the torque load on the engine. Thus as the aircraft speeds up (as in a dive) or slows (in climb) the RPM is held constant.

Small Engines

Small engines, such as used to power lawn mowers, portable generators, and lawn and garden tractors, are equipped with a governor to fuel limit the engine to a maximum safe speed when unloaded and to maintain a relatively constant speed despite changes in loading. In the case of generator applications, the engine speed must be closely controlled so the output frequency of the generator will remain reasonably constant.

Small engine governors are typically one of three types:

- **Pneumatic**: the governor mechanism detects air flow from the flywheel blower used to cool an air-cooled engine. The typical design includes an air vane mounted inside the engine's blower housing and linked to the carburetor's throttle shaft. A spring pulls the throttle open and, as the engine gains speed, increased air flow from the blower forces the vane back against the spring, partially closing the throttle. Eventually a point of equilibrium will be reached and the engine will run at a relatively constant speed. Pneumatic governors are simple in design and inexpensive to produce. However, they do not regulate engine speed very accurately and are affected by air density, as well as external conditions that may influence airflow.

- **Centrifugal**: a flyweight mechanism driven by the engine is linked to the throttle and works against a spring in a fashion similar to that of the pneumatic governor, resulting in essentially identical operation.

A centrifugal governor is more complex to design and produce than a pneumatic governor. However, the centrifugal design is more sensitive to speed changes and hence is better suited to engines that experience large fluctuations in loading.

- **Electronic**: a servo motor is linked to the throttle and controlled by an electronic module that senses engine speed by counting electrical pulses emitted by the ignition system or a magnetic pickup. The frequency of these pulses varies directly with engine speed, allowing the control module to apply a proportional voltage to the servo to regulate engine speed. Due to their sensitivity and rapid response to speed changes, electronic governors are often fitted to engine-driven generators designed to power computer hardware, as the generator's output frequency must be held within narrow limits to avoid malfunction.

Turbine Controls

In steam turbines, the steam turbine governing is the procedure of monitoring and controlling the flow rate of steam into the turbine with the objective of maintaining its speed of rotation as constant. The flow rate of steam is monitored and controlled by interposing valves between the boiler and the turbine.

In water turbines, the governors have been used since the mid-19th century to control the speeds of the turbines. A variety of flyball systems were used during the first 100 years of water turbine governors. Flyball component acted directly to the value of the turbine or the wicket gate to control the amount of water that enters the turbines. A newer system with mechanical governors started around 1880. An early mechanical governors is a servomechanism that comprises a series of gears that use the turbine's speed to drive the flyball and turbine's power to drive the control mechanism. The mechanical governors were continued to be enhanced. By 1930, the mechanical governors had many parameters that could be set on the feedback system for precise controls. In the later part of the twentieth century, electronic governors and digital systems started to replace the mechanical governors.

Other Uses of the Term

Computing

The Linux kernel has a number of *CPU frequency governors*, which are a sort of policies that set the CPU frequency based on the selected governor and usage patterns. For example, when the "performance" governor is active, the CPU frequency will be set to its maximum value, the "powersave" governor sets the CPU to its lowest frequency, the "ondemand" governor sets the CPU frequency depending on the current usage, *etc.*

CURRENT LOOP

In electrical signalling an analog **current loop** is used where a device must be monitored or controlled remotely over a pair of conductors. Only one current level can be present at any time.

Given its analog nature, current loops are easier to understand and debug than more complicated digital fieldbuses, requiring only a handheld digital multimeter in most situations. Using fieldbuses and solving related problems usually requires much more education and understanding than required by simple current loop systems.

Additional digital communication to the device can be added to current loop using HART Protocol. Digital process buses such as FOUNDATION Fieldbus and Profibus may replace analog current loops.

Process-control Use

For industrial process control instruments, analog 4–20 mA and 10–50 mA current loops are commonly used for analog signaling, with 4 mA representing the lowest end of the range and 20 mA the highest. The key advantages of the current loop are that the accuracy of the signal is not affected by voltage drop in the interconnecting wiring, and that the loop can supply operating power to the device. Even if there is significant electrical resistance in the line, the current loop transmitter will maintain the proper current, up to its maximum voltage capability.

The *live-zero* represented by 4 mA allows the receiving instrument to detect some failures of the loop, and also allows transmitter devices to be powered by the same current loop (called *two-wire* transmitters). Such instruments are used to measure pressure, temperature, level, flow, pH or other process variables. A current loop can also be used to control a valve positioner or other output actuator. An analog current loop can be converted to a voltage input with a precision resistor. Since input terminals of instruments may have one side of the current loop input tied to the chassis ground (earth), analog isolators may be required when connecting several instruments in series.

Depending on the source of current for the loop, devices may be classified as *active* (supplying power) or *passive* (relying on loop power). For example, a chart recorder may provide loop power to a pressure transmitter. The pressure transmitter modulates the current on the loop to send the signal to the strip chart recorder, but does not in itself supply power to the loop and so is passive. (A *4-wire* instrument has a power supply input separate from the current loop.) Another loop may contain two passive chart recorders, a passive pressure transmitter, and a 24 V battery. (The battery is the active device).

Panel mount displays and chart recorders are commonly termed 'indicator devices' or 'process monitors'. Several passive indicator devices may be connected in series, but a loop must have only one transmitter device and only one power source (active device).

The relationship between current value and process variable measurement is set by calibration, which assigns different ranges of engineering units to the span between 4 and 20 mA. The mapping between engineering units and current can be inverted, so that 4 mA represents the maximum and 20 mA the minimum.

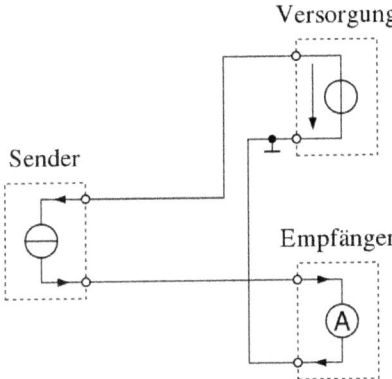

Fig. : Typ 2

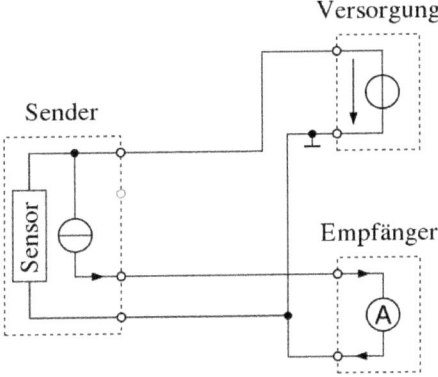

Fig. : Typ 3

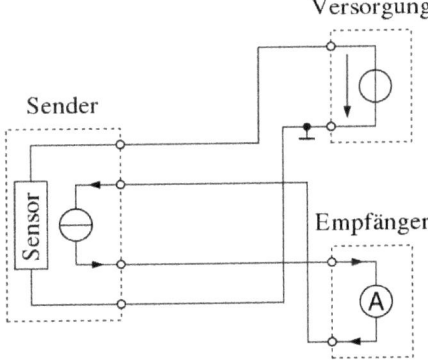

Fig. : Typ 4

Long Circuits

Analog current loops were occasionally carried between buildings by dry pairs in telephone cables leased from the local telephone company. 4–20 mA loops were more common in the days of analog telephony. These circuits require end-to-end direct current (DC) continuity. DC continuity is not available over a microwave radio, optical fibre, or a multiplexed telephone circuit connection.

Basic DC circuit theory shows that the current is the same all along the line. It was common to see 4–20 mA circuits that had loop lengths in miles or circuits working over telephone cable pairs that were longer than ten thousand feet end-to-end. There are still legacy systems in place using this technology. In Bell System circuits, voltages up to 125 VDC were employed.

Discrete Control

Discrete control functions can be represented by discrete levels of current sent over a loop. This would allow multiple control functions to be operated over a single pair of wires. Currents required for a specific function vary from one application or manufacturer to another. There is no specific current that is tied to a single meaning. It is almost universal that 0 mA indicates the circuit has failed. In the case of a fire alarm, 6 mA could be normal, 15 mA could mean a fire has been detected, and 0 mA would produce a trouble indication, telling the monitoring site the alarm circuit had failed. Some devices, such as two-way radio remote control consoles, can reverse the polarity of currents and can multiplex audio onto a DC current.

These devices can be employed for any remote control need a designer might imagine. For example, a current loop could actuate an evacuation siren or command synchronized traffic signals.

Two-way Radio Use

Current loop circuits are one possible way used to control radio base stations at distant sites. The two-way radio industry calls this type of remote control **DC remote**. This name comes from the need for DC circuit continuity between the control point and the radio base station. A current loop remote control saves the cost of extra pairs of wires between the operating point and the radio transceiver. Some equipment, such as the Motorola MSF-5000 base station, uses currents below 4 mA for some functions. An alternative type, the tone remote, is more complex but requires only an audio path between control point and base station. The patent does not describe this tone remote but confirms the use of the phrase to describe this system of signaling.

For example, a taxi dispatch base station might be physically located on the rooftop of an eight-story building. The taxi company office might be in the basement of a different building nearby. The office would have a remote control unit that would operate the taxi company base station over a current loop circuit. The circuit would normally be over a telephone line or similar wiring. Control function currents come from the remote control console at the dispatch office end of a circuit. In two-way radio use, an idle circuit would normally have no current present.

In two-way radio use, radio manufacturers use different currents for specific functions. Polarities are changed to get more possible functions over a single circuit. For example, imagine one possible scheme where the presence of these currents cause the base station to change state:

- no current means *receive on channel 1*, (the default).
- +6 mA might mean *transmit on channel 1*
- −6 mA might mean *stay in receive mode but switch to channel 2*. So long as the −6 mA current were present, the remote base station would continue to receive on channel 2.
- −12 mA might command the base station to *transmit on channel 2*.

This circuit is polarity-sensitive. If a telephone company cable splicer accidentally reversed the conductors, selecting channel 2 would lock the transmitter on.

Each current level could close a set of contacts, or operate solid-state logic, at the other end of the circuit. That contact closure caused a change of state on the controlled device. Some remote control equipment could have options set to allow compatibility between manufacturers. That is, a base station that was configured to transmit with a +18 mA current could have options changed to (instead) make it transmit when +6 mA was present.

In two-way radio use, AC signals were also present on the circuit pair. If the base station were idle, receive audio would be sent over the line from the base station to the dispatch office. In the presence of a transmit command current, the remote control console would send audio to be transmitted. The voice of the user in the dispatch office would be modulated and superimposed over the DC current that caused the transmitter to operate.

CRUISE CONTROL

Cruise control (sometimes known as speed control or autocruise, or tempomat in some countries) is a system that automatically controls the speed of a motor vehicle. The system is a servomechanism that takes over the throttle of the car to maintain a steady speed as set by the driver.

Speed control with a centrifugal governor was used in automobiles as early as 1900 in the Wilson-Pilcher and also in the 1910s by Peerless. Peerless advertised that their system would "maintain speed whether up hill or down". The technology was adopted by James Watt and Matthew Boulton in 1788 to control steam engines, but the use of governors dates at least back to the 17th century. On an engine the governor adjusts the throttle position as the speed of the engine changes with different loads, so as to maintain a near constant speed.

Fig. : Cruise control mounted on a 2000 Jeep Grand Cherokee steering wheel.

Modern cruise control (also known as a speedostat or tempomat) was invented in 1948 by the inventor and mechanical engineer Ralph Teetor. His idea was born out of the frustration of riding in a car driven by his lawyer, who kept speeding up and slowing down as he talked. The first car with Teetor's system was the 1958 Imperial (called "Auto-pilot") using a speed dial on the dashboard. This system calculated ground speed based on driveshaft rotations off the rotating speedometer-cable, and used a bi-directional screw-drive electric motor to vary throttle position as needed.

A 1955 U.S. Patent for a "Constant Speed Regulator" was filed in 1950 by M-Sgt Frank J. Riley. He installed his invention, which he conceived while driving on the Pennsylvania Turnpike, on his own car in 1948. Despite this patent, the inventor, Riley, and the subsequent patent holders were not able to collect royalties for any of the inventions using cruise control.

In 1965, American Motors (AMC) introduced a low-priced automatic speed control for its large-sized cars with automatic transmissions. The AMC "Cruise-Command" unit was engaged by a push-button once the desired speed was reached and then the throttle position was by a vacuum control directly from the speedometer cable rather than a separate dial on the dashboard.

Daniel Aaron Wisner invented "Automotive Electronic Cruise Control" in 1968 as an engineer for RCA's Industrial and Automation Systems Division in Plymouth, Michigan. His invention described in two patents filed that year (US 3570622 & US 3511329), with the second modifying his original design by debuting digital memory, was the first electronic device in controlling a car. Two decades passed before an integrated circuit for his design was developed by Motorola. as the MC14460 Auto Speed Control Processor in CMOS. The advantage of electronic speed control over its mechanical predecessor was that it could be integrated with electronic accident avoidance and engine management systems.

Following the 1973 oil crisis and rising fuel prices, the device became more popular in the U.S. "Cruise control can save gas by avoiding surges that expel fuel" while driving at steady speeds. In 1974, AMC, GM, and Chrysler priced the option at $60 to $70, while Ford charged $103.

Operation

The driver must bring the vehicle up to speed manually and use a button to set the cruise control to the current speed.

The cruise control takes its speed signal from a rotating driveshaft, speedometer cable, wheel speed sensor from the engine's RPM, or from internal speed pulses produced electronically by the vehicle. Most systems do not allow the use of the cruise control below a certain speed - typically around 25 mph (40 km/h). The vehicle will maintain the desired speed by pulling the throttle cable with a solenoid, a vacuum driven servomechanism, or by using the electronic systems built into the vehicle (fully electronic) if it uses a 'drive-by-wire' system.

All cruise control systems must be capable of being turned off both explicitly and automatically when the driver depresses the brake, and often also the clutch. Cruise control often includes a memory feature to resume the set speed after braking, and a coast feature to reduce the set speed without braking. When the cruise control is engaged, the throttle can still be used to accelerate the car, but once the pedal is released the car will then slow down until it reaches the previously set speed.

On the latest vehicles fitted with electronic throttle control, cruise control can be easily integrated into the vehicle's engine management system. Modern "adaptive" systems (see below) include the ability to automatically reduce speed when the distance to a car in front, or the speed limit, decreases. This is an advantage for those driving in unfamiliar areas.

The cruise control systems of some vehicles incorporate a "speed limiter" function, which will not allow the vehicle to accelerate beyond a pre-set maximum; this can usually be overridden by fully depressing the accelerator pedal. (Most systems will prevent the vehicle accelerating beyond the chosen speed, but will not apply the brakes in the event of overspeeding downhill.)

On vehicles with a manual transmission, cruise control is less flexible because the act of depressing the clutch pedal and shifting gears usually disengages the cruise control. The "resume" feature has to be used each time after selecting the new gear and releasing the clutch. Therefore, cruise control is of most benefit at motorway/highway speeds when top gear is used virtually all the time.

Advantages and Disadvantages

Some advantages of cruise control include:

- Its usefulness for long drives (reducing driver fatigue, improving comfort by allowing positioning changes more safely) across highways and sparsely populated roads.

- Some drivers use it to avoid subconsciously violating speed limits. A driver who otherwise tends to subconsciously increase speed over the course of a highway journey may avoid speeding.

However, when used incorrectly cruise control can lead to accidents due to several factors, such as:

- Driving over "rolling" terrain, with gentle up and down portions, can usually be done more economically (using less fuel) by a skilled driver viewing the approaching terrain, by maintaining a relatively constant throttle position and allowing the vehicle to accelerate on the downgrades and decelerate on upgrades, while reducing power when cresting a rise and adding a bit before an upgrade is reached. Cruise control will tend to overthrottle on the upgrades and retard on the downgrades, wasting the energy storage capabilities available from the inertia of the vehicle.

Adaptive Cruise Control

Some modern vehicles have **adaptive cruise control (ACC)** systems, which is a general term meaning improved cruise control. These improvements can be automatic braking or dynamic set-speed type controls.

Automatic Braking Type: The automatic braking type use either a radar or laser setup to allow the vehicle to keep pace with the car it is following, slow when closing in on the vehicle in front and accelerating again to the preset speed when traffic allows. Some systems also feature forward collision warning systems, which warns the driver if a vehicle in front — given the speed of both vehicles — gets too close (within the preset headway or braking distance).

Dynamic Set Speed Type: The dynamic set speed uses the GPS position of speed limit signs, from a database. Some are modifiable by the driver.

Non-Braking Type: The speed can be adjusted to allow traffic calming. One visual method uses OpenCV

CHECKWEIGHER

A **checkweigher** is an automatic or manual machine for checking the weight of packaged commodities. It is normally found at the offgoing end of a production process and is used to ensure that the weight of a pack of the commodity is within specified limits. Any packs that are outside the tolerance are taken out of line automatically.

A checkweigher can weigh in excess of 500 items per minute (depending on carton size and accuracy requirements). Checkweighers often incorporate additional checking devices such as metal detectors and X-ray machines to enable other attributes of the pack to be checked and acted upon accordingly.

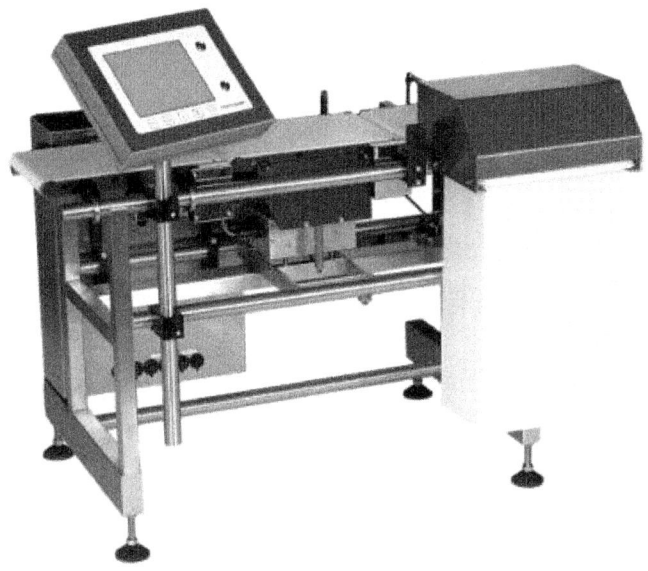

Fig. : Example checkweigher. Product passes on the conveyor belt where it is weighed.

A Typical Machine

An automatic checkweigher incorporates a series of conveyor belts. These checkweighers are known also as belt weighers, in-motion scales, conveyor scales, dynamic scales, and in-line scales. In filler applications, they are known as check scales. Typically, there are three belts or chain beds:

- An infeed belt that may change the speed of the package and to bring it up or down to a speed required for weighing. The infeed is also sometimes used as an indexer, which sets the gap between products to an optimal distance for weighing. It sometimes has special belts or chains to position the product for weighing.

- A weigh belt. This is typically mounted on a weight transducer which can typically be a strain-gauge load cell or a servo-balance (also known as a force-balance), or sometimes known as a split-beam. Some older machines may pause the weigh bed belt before taking the weight measurement. This may limit line speed and throughput.

- A reject belt that provides a method of removing an out-of-tolerance package from the conveyor line. The reject can vary by application. Some require an air-amplifier to blow small products off the belt, but heavier applications require a linear or radial actuator. Some fragile products are rejected by "dropping" the bed so that the product can slide gently into a bin or other conveyor.

For high-speed precision scales, a load cell using electromagnetic force restoration (EMFR) is appropriate. This kind of system charges an inductive coil, effectively floating the weigh bed in an electromagnetic field. When the weight is added, the movement of a ferrous material through that coil causes a fluctuation in the coil current proportional to the weight of the object. Other technologies used include strain gauges and vibrating wire load cells.

It is usual for a built-in computer to take many weight readings from the transducer over the time that the package is on the weigh bed to ensure an accurate weight reading.

Calibration is critical. A lab scale, which usually is in an isolated chamber pressurized with dry nitrogen(pressurized at sea level) can weigh an object within plus or minus 100th of a gram, but ambient air pressure is a factor. This is straightforward when there is no motion, but in motion there is a factor that is not obvious-noise from the motion of a weigh belt, vibration, air-conditioning or refrigeration which can cause drafts. Torque on a load cell causes erratic readings.

A dynamic, in-motion checkweigher takes samples, and analyzes them to form an accurate weight over a given time period. In most cases, there is a trigger from an optical(or ultrasonic) device to signal the passing of a package. Once the trigger fires, there is a delay set to allow the package to move to the "sweet spot" (center) of the weigh bed to sample the weight. The weight is sampled for a given duration. If either of these times are wrong, the weight will be wrong. There seems to be no scientific method to predict these timings. Some systems have a "graphing" feature to do this, but it is generally more of an empirical method that works best.

- A reject conveyor to enable the out-of-tolerance packages to be removed from the normal flow while still moving at the conveyor velocity. The reject mechanism can be one of several types. Among these are a simple pneumatic pusher to push the reject pack sideways from the belt, a diverting arm to sweep the pack sideways and a reject belt that lowers or lifts to divert the pack vertically. A typical checkweigher usually has a bin to collect the out-of-tolerance packs. Sometimes these bins are provided with a lock, to prevent that out of specification items are fed back on the conveyor belt.

Tolerance Methods

There are several tolerance methods:

- The traditional "minimum weight" system where weights below a specified weight are rejected. Normally the minimum weight is the weight that is printed on the pack or a weight level that exceeds that to allow for weight losses after production such as evaporation of commodities that have a moisture content. The larger wholesale companies have mandated that any product shipped to them have accurate weight checks such that a customer can be confident that they are getting the amount of product for which they paid. These wholesalers charge large fees for inaccurately filled packages.

- The European Average Weight System which follows three specified rules known as the "Packers Rules".
- Other published standards and regulations such as NIST Handbook 133

Data Collection

There is also a requirement under the European Average Weight System that data collected by checkweighers is archived and is available for inspection. Most modern checkweighers are therefore equipped with communications ports to enable the actual pack weights and derived data to be uploaded to a host computer. This data can also be used for management information enabling processes to be fine-tuned and production performance monitored.

Checkweighers that are equipped with high speed communications such as Ethernet ports are capable of integrating themselves into groups such that a group of production lines that are producing identical products can be considered as one production line for the purposes of weight control. For example, a line that is running with a low average weight can be complemented by another that is running with a high average weight such that the aggregate of the two lines will still comply with rules.

An alternative is to program the checkweigher to check bands of different weight tolerances. For instance, the total valid weight is 100 grams ±15 grams. This means that the product can weigh 85 g - 115 g. However, it is obvious that if you are producing 10,000 packs a day, and most of your packs are 110 g, you are losing 100 kg of product. If you try to run closer to 85 g, you may have a high rejection rate.

Example: A checkweigher is programmed to indicate 5 zones with resolution to 1 g:

1. Under Reject.... the product weighs 84.9 g or less
2. Under OK........ the product weighs 85 g, but less than 95 g
3. Valid........... the product weighs 96 g, but less than 105 g
4. Over OK......... the product weighs 105 g, and less than 114 g
5. Over Reject..... the product weighs over the 115 g limit

With a check weigher programmed as a zone checkweigher, the data collection over the networks, as well as local statistics, can indicate the need to check the settings on the upstream equipment to better control flow into the packaging. In some cases the dynamic scale sends a signal to a filler, for instance, in real-time, controlling the actual flow into a barrel, can, bag, *etc.* In many cases a checkweigher has a light-tree with different lights to indicate the variation of the zone weight of each product.

Application Considerations

Speed and accuracy that can be achieved by a checkweigher is influenced by the following:

- Pack length
- Pack weight
- Line speed required
- Pack content (solid or liquid)
- Motor technology
- Stabilization time of the weight transducer
- Airflow causing readings in error
- Vibrations from machinery causing unnecessary rejects
- Sensitivity to temperature, as the load cells *can* be temperature sensitive

Applications

In-motion scales are dynamic machines that can be designed to perform thousands of tasks. Some are used as simple caseweighers at the end of the conveyor line to ensure the overall finished package product is within its target weight.

An in motion conveyor checkweigher can be used to detect missing pieces of a kit, such as a cell phone package that is missing the manual, or other collateral. Checkweighers are typically used on the incoming conveyor chain, and the output pre-packaging conveyor chain in a poultry processing plant. The bird is weighed when it comes onto the conveyor, then after processing and washing at the end, the network computer can determine whether or not the bird absorbed too much water, which as it is further processed, will be drained, making the bird under its target weight.

A high speed conveyor scale can be used to change the pacing, or pitch of the products on the line by speeding, or slowing the product speed to change the distance between packs before reaching a different speed going into a conveyor machine that is boxing multiple packs into a box. The "pitch" is the measurement of the product as it comes down the conveyor line from leading edge to leading edge.

A checkweigher can be used to count packs, and the aggregate (total) weight of the boxes going onto a pallet for shipment, including the ability to read each package's weight and cubic dimensions. The controller computer can print a shipping label and a bar-code label to identify the weight, the cubic dimensions, ship-to address, and other data for machine ID through the shipment of the product. A receiving checkweigher for the shipment can read the label with a bar code scanner, and determine if the shipment is as it was before the transportation carrier received it from the shipper's loading dock, and determine if a box is missing, or something was pilfered or broken in transit.

Checkweighers are also used for Quality management. For instance, raw material for machining a bearing is weighed prior to beginning the process, and after the process, the quality inspector expects that a certain amount of metal was removed in the finishing process. The finished bearings are checkweighed, and bearings over- or underweight are rejected for physical inspection. This is a benefit to the inspector, since he can have a high confidence that the ones not rejected are within machining tolerance. A common usage is for throttling plastic extruders such that a bottle used to package detergent meets that requirements of the finished packager.

Quality management can use a checkweigher for Nondestructive testing to verify finished goods using common Evaluation methods to detect pieces missing from a "finished" product, such as grease from a bearing, or a missing roller within the housing.

Checkweighers can be built with metal detectors, x-ray machines, open-flap detection, bar-code scanners, holographic scanners, temperature sensors, vision inspectors, timing screws to set the timing and spacing between product, indexing gates and concentrator ducts to line up the product into a designated area on the conveyor. An industrial motion checkweigher can sort products from a fraction of a gram to many, many kilograms. In English units, is this from less than 100th of an ounce to as much as 500 lbs or more. Specialized checkweighers can weigh commercial aircraft, and even find their center-of-gravity.

Checkweighers can operate at very high speeds, processing products weighing fractions of a gram at over 100m/m (meters per minute) and materials such as pharmaceuticals and 200 lb bags of produce at over 100fpm(feet per minute). They can be designed in many shapes and sizes, hung from ceilings, raised on mezzanines, operated in ovens or in refrigerators. Their conveying medium can be industrial belting, low-static belting, chains similar to bicycle chains(but much smaller), or interlocked chain belts of any width. They can have chain belts made of special materials, different polymers, metals, *etc.*

Checkweighers are used in cleanrooms, dry atmosphere environments, wet environments, produce barns, food processing, drug processing, *etc.* Checkweighers are specified by the kind of environment, and the kind of cleaning will be used. Typically, a checkweigher for produce is made of mild steel, and one that will be cleaned with harsh chemicals, such as bleach, will be made with all stainless steel parts, even the Load cells. These machines are labeled "full washdown", and must have every part and component specified to survive the washdown environment.

Checkweighers are operated in some applications for extremely long periods of time- 24/7 year round. Generally, conveyor lines are not stopped unless there is maintenance required, or there is an emergency stop, called an E-stop. Checkweighers operating in high density conveyor lines may have numerous special equipments in their design to ensure that if an E-stop occurs, all power going to all motors is removed until the E-stop is cleared and reset.

FEEDBACK

Feedback occurs when outputs of a system are routed back as inputs as part of a chain of cause-and-effect that forms a circuit or loop. The system can then be said to *feed back* into itself.

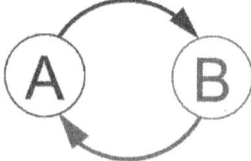

Fig. : Feedback exists between two parts when each affects the other.

The notion of cause-and-effect has to be handled carefully when applied to feedback systems:

"Simple causal reasoning about a feedback system is difficult because the first system influences the second and second system influences the first, leading to a circular argument. This makes reasoning based upon cause and effect tricky, and it is necessary to analyze the system as a whole."

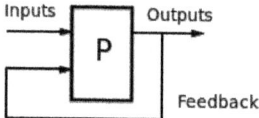

Fig. : A feedback loop where all outputs of a process are available as causal inputs to that process.

Self-regulating mechanisms have existed since antiquity, and the idea of feedback had started to enter economic theory in Britain by the eighteenth century, but it wasn't at that time recognized as a universal abstraction and so didn't have a name.

The verb phrase "to feed back", in the sense of *returning to an earlier position* in a mechanical process, was in use in the US by the 1860s, and in 1909, Nobel laureate Karl Ferdinand Braun used the term "feed-back" as a noun to refer to (undesired) *coupling* between components of an electronic circuit.

By the end of 1912, researchers using early electronic amplifiers (audions) had discovered that deliberately coupling part of the output signal back to the input circuit would boost the amplification (through regeneration), but would also cause the audion to howl or sing. This action of feeding back of the signal from output to input gave rise to the use of the term "feedback" as a distinct word by 1920.

Over the years there has been some dispute as to the best definition of feedback. According to Ashby (1956), mathematicians and theorists interested in the *principles* of feedback mechanisms prefer the definition of *circularity of action*, which keeps the theory simple and consistent. For those with more *practical* aims, feedback should be a deliberate effect via some more tangible connection.

"[Practical experimenters] object to the mathematician's definition, pointing out that this would force them to say that feedback was present in the ordinary pendulum ... between its position and its momentum—a 'feedback' that, from the practical point of view, is somewhat mystical. To this the mathematician retorts that if feedback is to be considered present only when there is an actual wire or nerve to represent it, then the theory becomes chaotic and riddled with irrelevancies."

Focusing on uses in management theory, Ramaprasad (1983) defines feedback generally as "...information about the gap between the actual level and the reference level of a system parameter" that is used to "alter the gap in some way." He emphasizes that the information by itself is not feedback unless translated into action.

Types

Positive and Negative Feedback

There are two types of feedback: *positive feedback* and *negative feedback*.

As an example of negative feedback, the diagram might represent a cruise control system in a car, for example, that matches a target speed such as the speed limit. The controlled system is the car; its input includes the combined torque from the engine and from the changing slope of the road (the disturbance). The car's speed (status) is measured by a speedometer. The error signal is the departure of the speed as measured by the speedometer from the target speed (set point). This measured error is interpreted by the controller to adjust the accelerator, commanding the fuel flow to the engine (the effector). The resulting change in engine torque, the feedback, combines with the torque exerted by the changing road grade to reduce the error in speed, minimizing the road disturbance.

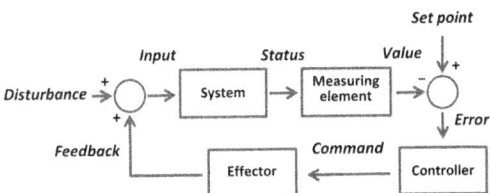

Fig. : Maintaining a desired system performance despite disturbance using negative feedback to reduce system error.

The terms "positive" and "negative" were first applied to feedback prior to WWII. The idea of positive feedback was already current in the 1920s with the introduction of the regenerative circuit. Friis and Jensen (1924) described regeneration in a set of electronic amplifiers as a case where the *"feed-back" action is positive* in contrast to negative feed-back action, which they mention only in passing. Harold Stephen Black's classic 1934 paper first details the use of negative feedback in electronic amplifiers. According to Black:

"Positive feed-back increases the gain of the amplifier, negative feed-back reduces it."

According to Mindell (2002) confusion in the terms arose shortly after this:

"...Friis and Jensen had made the same distinction Black used between 'positive feed-back' and 'negative feed-back', based not on the sign of the feedback itself but rather on its effect on the amplifier's gain. In contrast, Nyquist and Bode, when they built on Black's work, referred to negative feedback as that with the sign reversed. Black had trouble convincing others of the utility of his invention in part because confusion existed over basic matters of definition."

Even prior to the terms being applied, James Clerk Maxwell had described several kinds of "component motions" associated with the centrifugal governors used in steam engines, distinguishing between those that lead to a continual *increase* in a disturbance or the amplitude of an oscillation, and those that lead to a *decrease* of the same.

Terminology

The terms positive and negative feedback are defined in different ways within different disciplines.

1. The altering of the *gap* between reference and actual values of a parameter, based on whether the gap is *widening* (positive) or *narrowing* (negative).

2. The valence of the *action* or *effect* that alters the gap, based on whether it has a *happy* (positive) or *unhappy* (negative) emotional connotation to the recipient or observer.

The two definitions may cause confusion, such as when an incentive (reward) is used to boost poor performance (narrow a gap). Referring to definition 1, some authors use alternative terms, replacing *positive/negative* with *self-reinforcing/ self-correcting, reinforcing/balancing, discrepancy-enhancing/discrepancy-reducing* or *regenerative/degenerative* respectively. And for definition 2, some authors advocate describing the action or effect as positive/negative *reinforcement* or *punishment* rather than feedback. Yet even within a single discipline an example of feedback can be called either positive or negative, depending on how values are measured or referenced.

This confusion may arise because feedback can be used for either *informational* or *motivational* purposes, and often has both a *qualitative* and a *quantitative* component. As Connellan and Zemke (1993) put it:

"*Quantitative* feedback tells us how much and how many. *Qualitative* feedback tells us how good, bad or indifferent."

Limitations of Negative and Positive Feedback

While simple systems can sometimes be described as one or the other type, many systems with feedback loops cannot be so easily designated as simply positive or negative, and this is especially true when multiple loops are present.

"When there are only two parts joined so that each affects the other, the properties of the feedback give important and useful information about the properties of the whole. But when the parts rise to even as few as four, if every one affects the other three, then twenty circuits can be traced through them; and knowing the properties of all the twenty circuits does not give complete information about the system."

Other Types of Feedback

In general, feedback systems can have many signals fed back and the feedback loop frequently contain mixtures of positive and negative feedback where positive and negative feedback can dominate at different frequencies or different points in the state space of a system.

The term bipolar feedback has been coined to refer to biological systems where positive and negative feedback systems can interact, the output of one affecting the input of another, and vice versa.

Some systems with feedback can have very complex behaviors such as chaotic behaviors in non-linear systems, while others have much more predictable behaviors, such as those that are used to make and design digital systems.

Feedback is used extensively in digital systems. For example, binary counters and similar devices employ feedback where the current state and inputs are used to calculate a new state which is then fed back and clocked back into the device to update it.

Applications

Biology

In biological systems such as organisms, ecosystems, or the biosphere, most parameters must stay under control within a narrow range around a certain optimal level under certain environmental conditions. The deviation of the optimal value of the controlled parameter can result from the changes in internal and external environments. A change of some of the environmental conditions may also require change of that range to change for the system to function. The value of the parameter to maintain is recorded by a reception system and conveyed to a regulation module via an information channel. An example of this is Insulin oscillations.

Biological systems contain many types of regulatory circuits, both positive and negative. As in other contexts, *positive* and *negative* do not imply that the feedback causes *good* or *bad* effects. A negative feedback loop is one that tends to slow down a process, whereas the positive feedback loop tends to accelerate it. The mirror neurons are part of a social feedback system, when an observed action is "mirrored" by the brain—like a self-performed action.

Feedback is also central to the operations of genes and gene regulatory networks. Repressor and activator proteins are used to create genetic operons, which were identified by Francois Jacob and Jacques Monod in 1961 as *feedback loops*.

These feedback loops may be positive (as in the case of the coupling between a sugar molecule and the proteins that import sugar into a bacterial cell), or negative (as is often the case in metabolic consumption).

On a larger scale, feedback can have a stabilizing effect on animal populations even when profoundly affected by external changes, although time lags in feedback response can give rise to predator-prey cycles.

In zymology, feedback serves as regulation of activity of an enzyme by its direct product(s) or downstream metabolite(s) in the metabolic pathway.

The hypothalamic–pituitary–adrenal axis is largely controlled by positive and negative feedback, much of which is still unknown.

In psychology, the body receives a stimulus from the environment or internally that causes the release of hormones. Release of hormones then may cause more of those hormones to be released, causing a positive feedback loop. This cycle is also found in certain behaviour. For example, "shame loops" occur in people who blush easily. When they realize that they are blushing, they become even more embarrassed, which leads to further blushing, and so on.

Climate Science

The climate system is characterized by strong positive and negative feedback loops between processes that affect the state of the atmosphere, ocean, and land. A simple example is the ice-albedo positive feedback loop whereby melting snow exposes more dark ground (of lower albedo), which in turn absorbs heat and causes more snow to melt.

Control Theory

Feedback is extensively used in control theory, using a variety of methods including state space (controls), full state feedback (also known as pole placement), and so forth. Note that in the context of control theory, "feedback" is traditionally assumed to specify "negative feedback".

The most common general-purpose controller using a control-loop feedback mechanism is a proportional-integral-derivative (PID) controller. Heuristically, the terms of a PID controller can be interpreted as corresponding to time: the proportional term depends on the *present* error, the integral term on the accumulation of *past* errors, and the derivative term is a prediction of *future* error, based on current rate of change.

Mechanical Engineering

In ancient times, the float valve was used to regulate the flow of water in Greek and Roman water clocks; similar float valves are used to regulate fuel in a carburettor and also used to regulate tank water level in the flush toilet.

The Dutch inventor Cornelius Drebbel (1572-1633) built thermostats (c1620) to control the temperature of chicken incubators and chemical furnaces. In 1745, the windmill was improved by blacksmith Edmund Lee, who added a fantail to keep the face of the windmill pointing into the wind. In 1787, Thomas Mead regulated the rotation speed of a windmill by using a centrifugal pendulum to adjust the distance between the bedstone and the runner stone (*i.e.*, to adjust the load).

The use of the centrifugal governor by James Watt in 1788 to regulate the speed of his steam engine was one factor leading to the Industrial Revolution. Steam engines also use float valves and pressure release valves as mechanical regulation devices. A mathematical analysis of Watt's governor was done by James Clerk Maxwell in 1868.

The *Great Eastern* was one of the largest steamships of its time and employed a steam powered rudder with feedback mechanism designed in 1866 by John McFarlane Gray. Joseph Farcot coined the word *servo* in 1873 to describe steam-powered steering systems. Hydraulic servos were later used to position guns. Elmer Ambrose Sperry of the Sperry Corporation designed the first autopilot in 1912. Nicolas Minorsky published a theoretical analysis of automatic ship steering in 1922 and described the PID controller.

Internal combustion engines of the late 20th century employed mechanical feedback mechanisms such as the vacuum timing advance but mechanical feedback was replaced by electronic engine management systems once small, robust and powerful single-chip microcontrollers became affordable.

Electronic Engineering

The use of feedback is widespread in the design of electronic amplifiers, oscillators, and stateful logic circuit elements such as flip-flops and counters. Electronic feedback systems are also very commonly used to control mechanical, thermal and other physical processes.

If the signal is inverted on its way round the control loop, the system is said to have *negative feedback*; otherwise, the feedback is said to be *positive*. Negative feedback is often deliberately introduced to increase the stability and accuracy of a system by correcting or reducing the influence of unwanted changes. This scheme can fail if the input changes faster than the system can respond to it. When this happens, the lag in arrival of the correcting signal can result in over-correction, causing the output to oscillate or "hunt". While often an unwanted consequence of system behaviour, this effect is used deliberately in electronic oscillators.

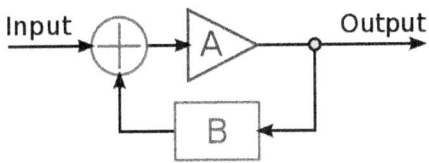

Fig. : The simplest form of a feedback amplifier can be represented by the ideal block diagram made up of unilateral elements.

Harry Nyquist contributed the Nyquist plot for assessing the stability of feedback systems. An easier assessment, but less general, is based upon gain margin and phase margin using Bode plots (contributed by Hendrik Bode). Design to ensure stability often involves frequency compensation, one method of compensation being pole splitting.

Electronic feedback loops are used to control the output of electronic devices, such as amplifiers. A feedback loop is created when all or some portion of the output is fed back to the input. A device is said to be operating *open loop* if no output feedback is being employed and *closed loop* if feedback is being used.

When two or more amplifiers are cross-coupled using positive feedback, complex behaviors can be created. These *multivibrators* are widely used and include:

- astable circuits, which act as oscillators

- monostable circuits, which can be pushed into a state, and will return to the stable state after some time

- bistable circuits, which have two stable states that the circuit can be switched between

Negative Feedback

Negative feedback occurs when the fed-back output signal has a relative phase of 180° with respect to the input signal (upside down). This situation is sometimes referred to as being *out of phase*, but that term also is used to indicate other phase separations, as in "90° out of phase". Negative feedback can be used to correct output errors or to desensitize a system to unwanted fluctuations. In feedback amplifiers, this correction is generally for waveform distortion reduction or to establish a specified gain level. A general expression for the gain of a negative feedback amplifier is the asymptotic gain model.

Positive Feedback

Positive feedback occurs when the fed-back signal is in phase with the input signal. Under certain gain conditions, positive feedback reinforces the input signal to the point where the output of the device oscillates between its maximum and minimum possible states. Positive feedback may also introduce hysteresis into a circuit. This can cause the circuit to ignore small signals and respond only to large ones. It is sometimes used to eliminate noise from a digital signal. Under some circumstances, positive feedback may cause a device to latch, *i.e.*, to reach a condition in which the output is locked to its maximum or minimum state. This fact is very widely used in digital electronics to make bistable circuits for volatile storage of information.

The loud squeals that sometimes occurs in audio systems, PA systems, and rock music are known as audio feedback. If a microphone is in front of a loudspeaker that it is connected to, sound that the microphone picks up comes out of the speaker, and is picked up by the microphone and re-amplified. If the loop gain is sufficient, howling or squealing at the maximum power of the amplifier is possible.

Oscillator

An electronic oscillator is an electronic circuit that produces a periodic, oscillating electronic signal, often a sine wave or a square wave. Oscillators convert direct current (DC) from a power supply to an alternating current signal. They are widely used in many electronic devices. Common examples of signals generated by oscillators include signals broadcast by radio and television transmitters, clock signals that regulate computers and quartz clocks, and the sounds produced by electronic beepers and video games.

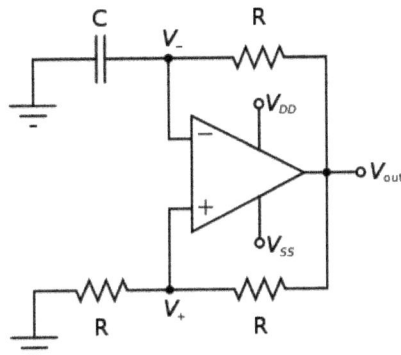

Fig. : A popular op-amp relaxation oscillator.

Oscillators are often characterized by the frequency of their output signal:

- A low-frequency oscillator (LFO) is an electronic oscillator that generates a frequency below ≈20 Hz. This term is typically used in the field of audio synthesizers, to distinguish it from an audio frequency oscillator.
- An audio oscillator produces frequencies in the audio range, about 16 Hz to 20 kHz.
- An RF oscillator produces signals in the radio frequency (RF) range of about 100 kHz to 100 GHz.

Oscillators designed to produce a high-power AC output from a DC supply are usually called inverters.

There are two main types of electronic oscillator: the linear or harmonic oscillator and the nonlinear or relaxation oscillator.

Latches and Flip-flops

A latch or a flip-flop is a circuit that has two stable states and can be used to store state information. They typically constructed using feedback that crosses over between two arms of the circuit, to provide the circuit with a state. The circuit can be made to change state by signals applied to one or more control inputs and will have one or two outputs. It is the basic storage element in sequential logic. Latches and flip-flops are fundamental building blocks of digital electronics systems used in computers, communications, and many other types of systems.

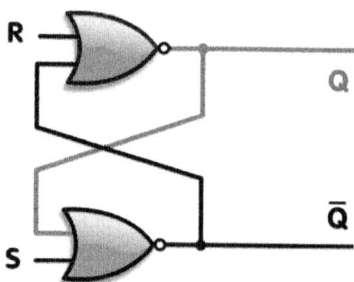

Fig. : An SR latch, constructed from a pair of cross-coupled NOR gates.

Latches and flip-flops are used as data storage elements. Such data storage can be used for storage of *state*, and such a circuit is described as sequential logic. When used in a finite-state machine, the output and next state depend not only on its current input, but also on its current state (and hence, previous inputs). It can also be used for counting of pulses, and for synchronizing variably-timed input signals to some reference timing signal.

Flip-flops can be either simple (transparent or opaque) or clocked (synchronous or edge-triggered). Although the term flip-flop has historically referred generically to both simple and clocked circuits, in modern usage it is common to reserve the term *flip-flop* exclusively for discussing clocked circuits; the simple ones are commonly called *latches*.

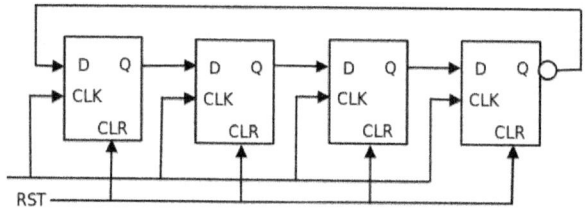

Fig. : A 4-bit ring counter using D-type flip flops.

Using this terminology, a latch is level-sensitive, whereas a flip-flop is edge-sensitive. That is, when a latch is enabled it becomes transparent, while a flip flop's output only changes on a single type (positive going or negative going) of clock edge.

Software

Feedback loops provide generic mechanisms for controlling the running, maintenance, and evolution of software and computing systems. Feedback-loops are important models in the engineering of adaptive software, as they define the behaviour of the interactions among the control elements over the adaptation process, to guarantee system properties at run-time. Feedback loops and foundations of control theory have been successfully applied to computing systems. In particular, they have been applied to the development of products such as IBM's Universal Database server and IBM Tivoli. From a software perspective, the autonomic (MAPE, monitor analyze plan execute) loop proposed by researchers of IBM is another valuable contribution to the application of feedback loops to the control of dynamic properties and the design and evolution of autonomic software systems.

Video Feedback

Video feedback is the video equivalent of acoustic feedback. It involves a loop between a video camera input and a video output, *e.g.*, a television screen or monitor. Aiming the camera at the display produces a complex video image based on the feedback.

Social Sciences

Economics and Finance

The stock market is an example of a system prone to oscillatory "hunting", governed by positive and negative feedback resulting from cognitive and emotional factors among market participants. For example,

- When stocks are rising (a bull market), the belief that further rises are probable gives investors an incentive to buy (positive feedback—reinforcing the rise, see also stock market bubble); but the increased price of the shares, and the knowledge that there must be a peak after which the market falls, ends up deterring buyers (negative feedback—stabilizing the rise).
- Once the market begins to fall regularly (a bear market), some investors may expect further losing days and refrain from buying (positive feedback—reinforcing the fall), but others may buy because stocks become more and more of a bargain (negative feedback—stabilizing the fall).

George Soros used the word *reflexivity*, to describe feedback in the financial markets and developed an investment theory based on this principle.

The conventional economic equilibrium model of supply and demand supports only ideal linear negative feedback and was heavily criticized by Paul Ormerod in his book *The Death of Economics*, which, in turn, was criticized by traditional economists. This book was part of a change of perspective as economists started to recognise that chaos theory applied to nonlinear feedback systems including financial markets.

Chapter 3

CONTROL LOOP ANALYSIS

Control theory is an interdisciplinary branch of engineering and mathematics that deals with the behavior of dynamical systems with inputs, and how their behavior is modified by feedback. The usual objective of control theory is to control a system, often called the *plant*, so its output follows a desired control signal, called the *reference*, which may be a fixed or changing value.

To do this a *controller* is designed, which monitors the output and compares it with the reference. The difference between actual and desired output, called the *error* signal, is applied as feedback to the input of the system, to bring the actual output closer to the reference. Some topics studied in control theory are stability (whether the output will converge to the reference value or oscillate about it), controllability and observability.

Extensive use is usually made of a diagrammatic style known as the block diagram. The transfer function, also known as the system function or network function, is a mathematical representation of the relation between the input and output based on the differential equations describing the system.

Although a major application of control theory is in control systems engineering, which deals with the design of process control systems for industry, other applications range far beyond this. As the general theory of feedback systems, control theory is useful wherever feedback occurs. A few examples are in physiology, electronics, climate modeling, machine design, ecosystems, navigation, neural networks, predator-prey interaction, gene expression, and production theory.

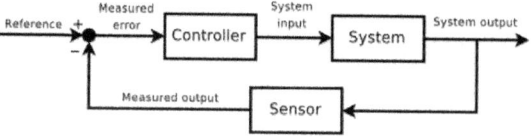

Fig. : The concept of the feedback loop to control the dynamic behavior of the system: this is negative feedback, because the sensed value is subtracted from the desired value to create the error signal, which is amplified by the controller.

Control theory is :

- a theory that deals with influencing the behavior of dynamical systems
- an interdisciplinary subfield of science, which originated in engineering and mathematics, and evolved into use by the social sciences, such as economics, psychology, sociology, criminology and in the financial system.

Control systems may be thought of as having four functions: measure, compare, compute and correct. These four functions are completed by five elements: detector, transducer, transmitter, controller and final control element. The measuring function is completed by the detector, transducer and transmitter. In practical applications these three elements are typically contained in one unit. A standard example of a measuring unit is a resistance thermometer.

The compare and compute functions are completed within the controller, which may be implemented electronically by proportional control, a PI controller, PID controller, bistable, hysteretic control or programmable logic controller. Older controller units have been mechanical, as in a centrifugal governor or a carburetor. The correct function is completed with a final control element. The final control element changes an input or output in the control system that affects the manipulated or controlled variable.

SIMPLE SIGNAL INJECTOR AIDS CONTROL-LOOP ANALYSIS

A signal-injection circuit for control-loop analysis is flat from dc to 200 kHz, isolated from chassis ground and easily constructed with a readily available instrumentation amplifier

Network analysis is a powerful and well-established method of characterizing and optimizing a control system. Unfortunately, making a successful measurement can be difficult and frustrating without the proper instrumentation. Having a good network analyzer is not enough. There must be a means for injecting a test signal into a closed loop over the frequency range of interest.

A common method of signal injection is to insert a 100-W resistor in the control loop, typically between the error amp and the plant, which is everything between the control output and the feedback input. For example, a buck converter plant consists of the sawtooth generator and comparator, the power transistor, the catch diode and the LC filter. The injection point must be between a low-impedance output and a high-impedance input. A transformer is then used to generate an ac test signal across the resistor. The reference signal is then measured at the plant input and the response is measured at the error-amplifier output.

Figure shows a typical control loop and two common locations for signal injection and measurement. It is very difficult to design a transformer that will provide a flat signal both at very low frequencies (< 1 Hz) and at higher frequencies (200 kHz). An engineer can forego this design and purchase a transformer, but the commercial versions still have the same frequency limitations, typically only operating over one or two decades. Electronic injection circuits exist but are expensive, around $1000. At least one open-source design exists according to published literature, but it suffers from chassis grounding issues.

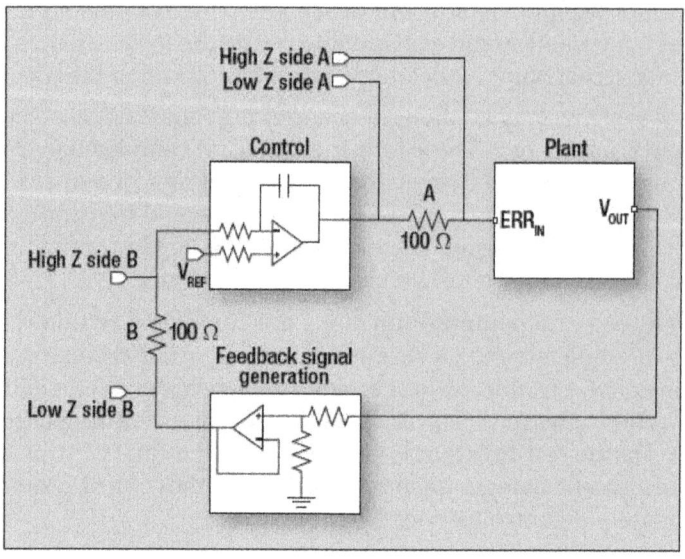

Fig. : In a typical control loop the error amplifier sends an error signal to the plant. Resistors A and B are possible points for signal injection.

Another Method

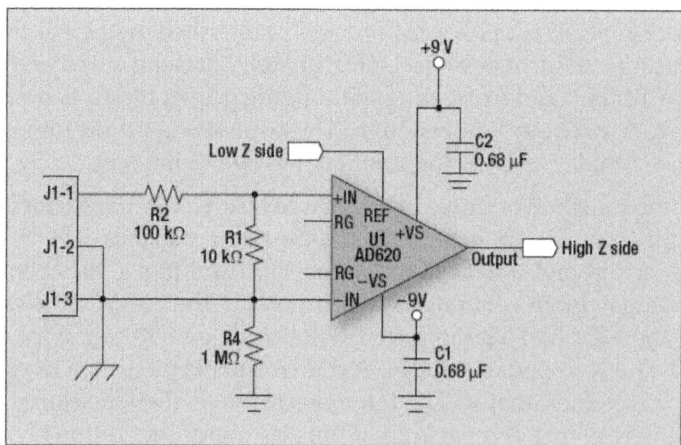

Fig. : In the instrumentation amplifier-based signal injection circuit the ouput of the error amp drives the "low Z side", and the "high Z side" drives the plant input.

Figure shows a simple circuit that performs the forementioned tasks very well. An instrumentation amplifier (inamp) isolates the sine-wave test signal from the network analyzer ground. The reference input is driven by the low-impedance output of an error amp or buffer. The sine wave then rides on the reference node, or control point of the circuit. It is important to realize that the reference input to an inamp is not a high-impedance node. Therefore, make sure that this node is truly driven by a low-impedance source.

The output of the inamp is connected to the high-impedance input of the plant. A designer does not actually need the 100-W resistor with this design, but its presence prevents accidental open-loop conditions if the device under test (DUT) is powered up before the test fixture. Analog Device's AD620 is able to be run on ±18-V rails, allowing for injection at nearly any conceivable point in most loops. Lastly, note the 1-MW resistor from the inverting input to ground. The inamp needs nanoamps of input bias current, which isn't much, but it is ill-advised to float the inputs completely. The resistor allows a very small amount of bias current without changing the return current paths appreciably.

Design Verification

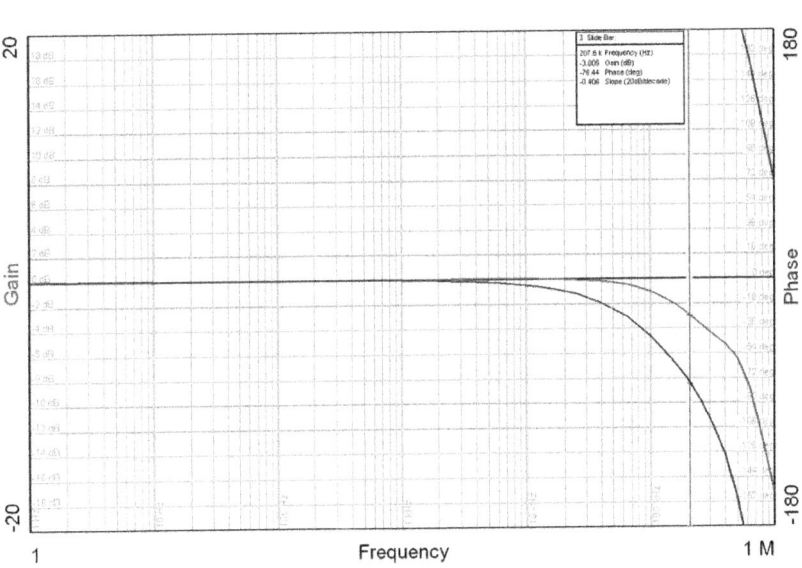

Bode Plot of Test Fixture Open Circuit

The injection circuit was tested using a Venable 3120 network analyzer. First, the open-circuit performance of the fixture was checked with a test voltage of 500 mV and the reference input biased at 1 V (a graph of open-circuit performance is shown as Figure) The region of practical use is from 0 kHz to 200 kHz. The phase shift is unimportant because it affects neither the network measurement nor the DUT. However, gain is important because the injected signal will start to decrease in the region of gain roll-off, reducing signal-to-noise and resulting in an inferior measurement. A higher-performance inamp would extend the range of the circuit.

It is very easy to believe erroneous results from a network analyzer. There are innumerable ways to compromise the measurement, from incorrect test signal amplitude to forgetting to turn on the injection circuit. It is important to have an oscilloscope hooked up to both the signal injection and measurement points. The signals should be relatively clean — distortion and noise compromise a reading. Once the setup has passed an open-loop test like the one previously described, it is prudent to test a standard, such as a simple RC circuit. As a means of verifying signal integrity, a buffered 10-kW/16-nF low-pass filter was tested in the fixture with a dc bias of 1 V.

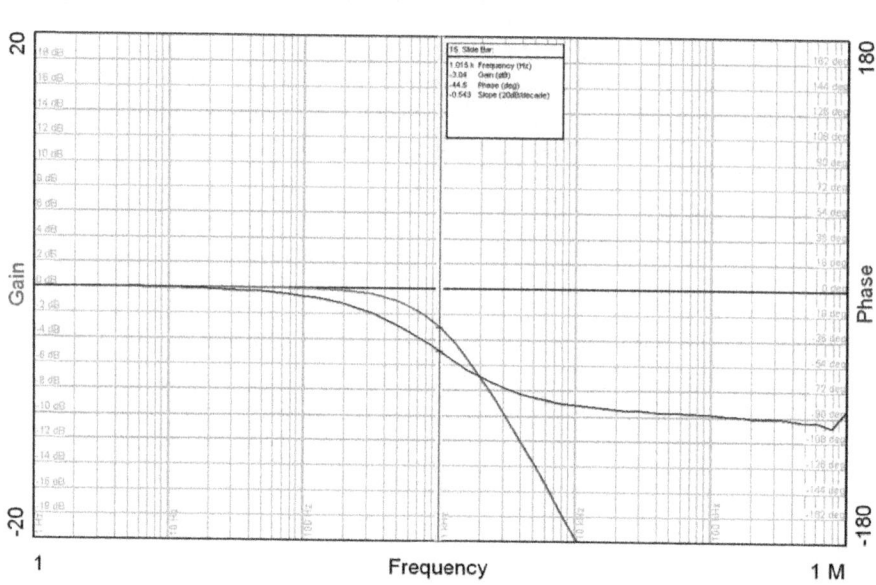

Bode Plot of 10k 16nF Low Pass Filter

Results of this test showed that the performance was very good, matching the expected 3-dB point of 1 kHz. There are many other logical test circuits, but making a standard that closely matches the application is best. The results from a 10-Hz high-pass filter won't yield much information about how the test circuit will behave at 100 kHz.

Plant Measurement

After verifying performance open loop and then in a test circuit, the test fixture was used to take practical data. The injection circuit was used to make actual network measurements of a 45-kV X-ray source's kilovolt-drive circuit. This particular drive is for a handheld X-ray fluorescence spectrometer used in RoHS compliance testing. Because it is battery powered, the spectrometer needs to be stable over a rather wide input range of 12 V to 18 V.

Load conditions also vary widely, from 10 kV to 45 kV and 0 mA to 50 mA. The initial "guess" compensation worked well over most line and load conditions, but the supply occasionally oscillated at high kilovolt and no load. To find out what was going wrong, the inamp test fixture was used to probe stability at several line and load combinations, including the problem region.

Even if the compensation is in place and working, it is usually very instructive to first look at the plant response. Run the system closed loop and place the test input at the input of the plant, but instead of placing the measurement input at the output of the error amp, place it at the output of the feedback generation circuit (or the output of the plant).

A 100-W resistor was inserted between the error amp and the plant, and the injector was connected as previously described. The reference input of the network analyzer was connected to the output of the inamp, and the signal input of the network analyzer was connected to the buffered feedback reference (1 V = 10 kV). The DUT was brought up to 45 kV and 40 mA of load current, and the injected sine-wave amplitude was adjusted to yield a good signal, but not so much as to disrupt the operating point (about 100 mV$_{PK-PK}$). A sweep from 1 Hz to 20 kHz yielded the results in Figure.

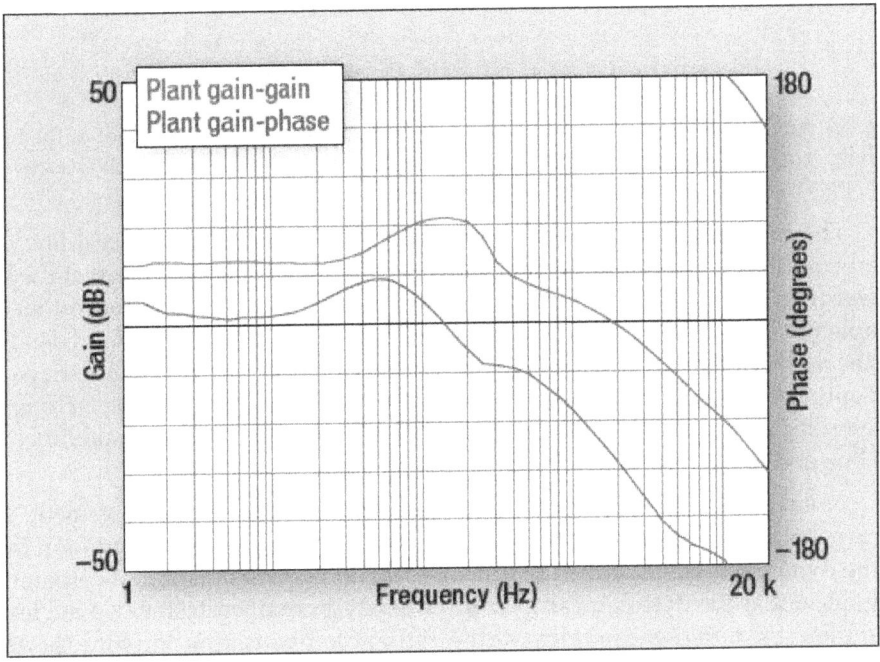

Fig.: This closed-loop Bode plot of the high-voltage plant was obtained by placing the reference node (S1) of the analyzer at "high Z side A" and the measurement node (S2) at "low Z side B."

By knowing that a type-1 integrating error amplifier provides a constant 270 degrees of phase shift (90 degrees from the capacitor and 180 degrees from the logic inversion of the amp), the phase can be shifted to see where the 0 degree crossing point is. The criterion for stability is that the gain must be less than 0 dB when the phase crosses 0 degrees. Knowing this, the integrator roll-off can be set for optimal performance.

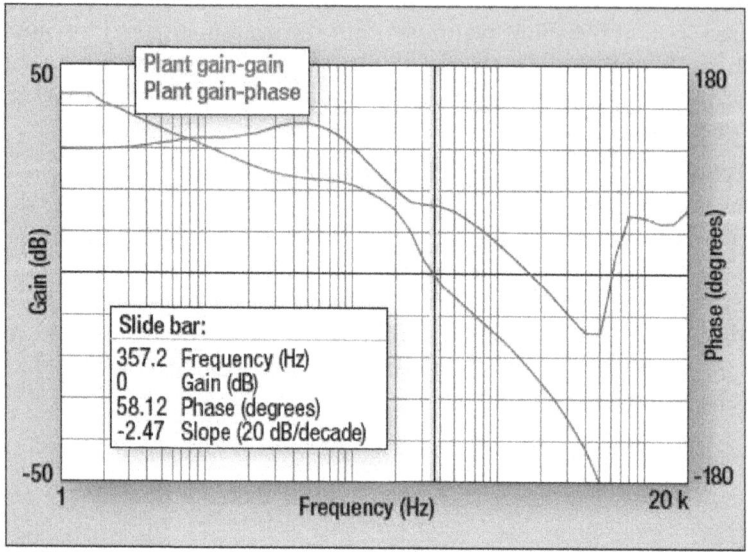

Fig. : A type-2 integrator was used to stabilize the loop in a typical Bode plot of the high-voltage supply's loop gain. Phase margin in this supply is 60 degrees, which ensures that the supply doesn't overshoot, but at the expense of settling time.

The measurement described previously was repeated for several line and load conditions. It was found that the dip in phase shown in Figure shifted to lower frequencies with decreasing load. This effect was so pronounced that the ample phase margin at 45 kV and 50 mA was reduced to zero at no load, leading to the oscillation described previously. H. Dean Venable wrote a landmark paper on optimizing feedback compensation networks. Through use of the techniques prescribed by Venable, the phase in that area was boosted to ensure stability over all line and load conditions.

Note that there are two practical ways to make this plant measurement. The loop usually can be stabilized by swamping the capacitance of the integrator. Even if the dynamic performance is abominable, it will keep the plant in the operating region so that the designer can perform the forementioned test. Once the test is complete, the designer can calculate the optimal compensation and tune the error amp to a stable, high-performance state.

A second way to perform this measurement is to run the plant open loop, using a variable voltage reference to set the operating point. This dc level is injected into the reference input of the inamp, and the output of the inamp drives the plant directly. On pulse-width modulation chips with internal amps, this node is often available as the COMP pin. Running the chip in this configuration can be a little tricky since ICs such as Linear Technology's LT3431 pack a bunch of features into the feedback node, for example frequency reduction and current limiting. Clamping the feedback input to the correct voltage will often solve these problems (*e.g.*, in the case of the LT3431, a voltage of 1 V keeps the chip operating normally).

Loop Measurement

The Bode plot in figure shows a plant gain of about 11 dB and a 90 degree phase shift at about 1.5 kHz. Using Venable's methods, the loop was compensated to have 60 degrees of phase margin (the phase at the 0-dB point). This results in a slightly overdamped response to step changes in program, which minimizes the voltage stress on the components. In other systems, fast settling time might be more important, which would indicate a lower phase margin, but resulting in more overshoot.

Figure shows the Bode plot of the compensated power supply at 45 kV and 50 mA. The results of the test match the predicted response of the measured plant gain combined with the integrator, indicating that the test fixture performs its task well.

Network analysis can be difficult not only because of the challenges of signal injection, but also due to the practical difficulties of maneuvering so many test leads and scope probes. It is very easy to short out a critical point and destroy equipment or the DUT.

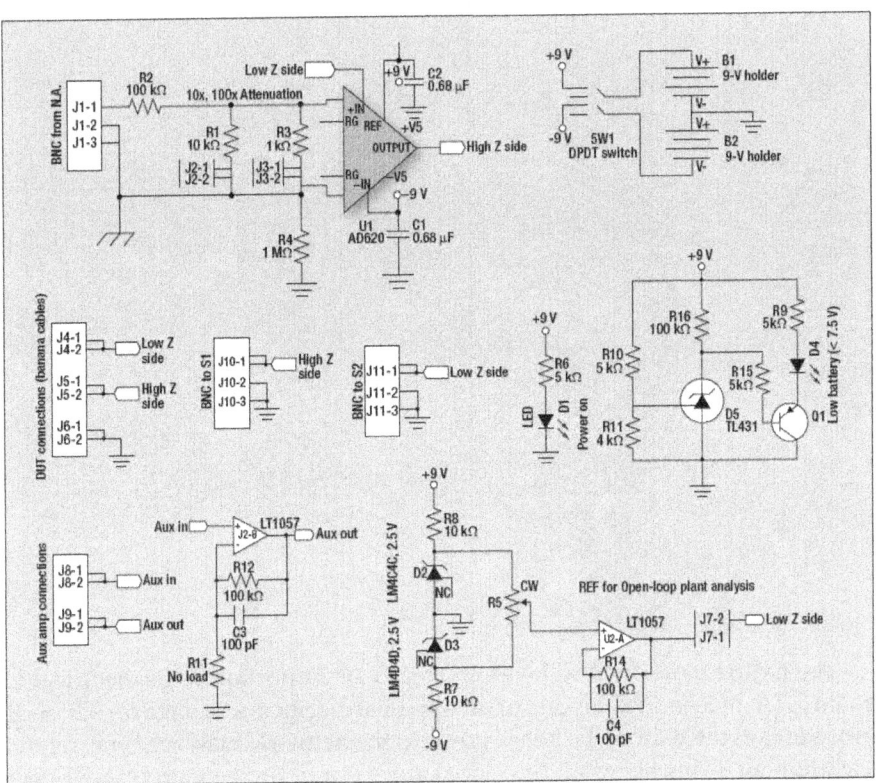

Fig. : Complete schematic of signal-injection pc board. The inamp in the upper left-hand corner does the actual signal injection.

Figure shows the full schematic of the signal injection circuit. In the upper left-hand corner, U1 handles the actual signal injection. Note that there is isolation from chassis ground at the signal input. R4 provides enough current for input bias current requirements, but shouldn't cause ground loop errors. Jumpers J2 and J3 provide 10x and 100x attenuation for fine control of the injected signal, which can be very helpful for analyzers with limited attenuation control. Two batteries provide low noise and isolated rails (this could be changed to extend the usable range to ±18 V).

In the lower right-hand corner, U2 (LT1057) provides buffering for a variable voltage reference (needed for making open-plant measurements) and an auxiliary buffer (for providing a low-impedance feedback point). To make the open-plant measurements, J7 is jumpered and the pot adjusted until the plant is at the desired operating point. The plant response is then measured from "high Z side A" to "low Z side B". The TL431 is used for a low-battery-detect circuit. The various connectors are used to interface between the DUT and the network analyzer.

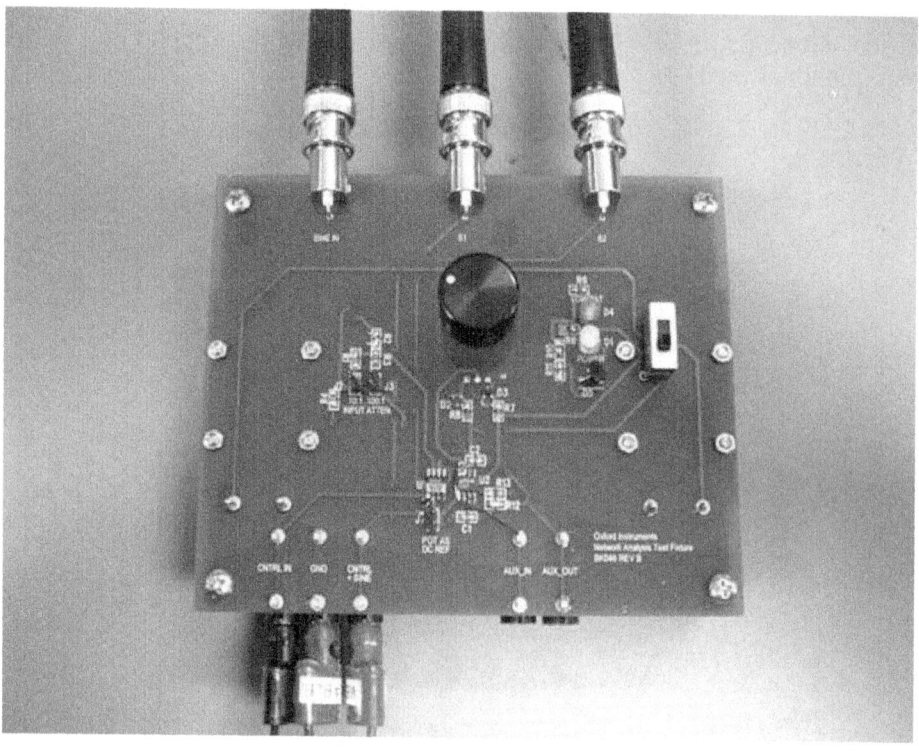

The fixture's mechanical layout is nearly as important as its electrical functionality. (A mechanical layout of the pc board appears as Figure) Three BNC connections at the back of the board go off to the network analyzer (sine, reference, measurement). This keeps all the large cables out of the way and under control. Two 9-V batteries (mounted on the underside of the board) provide power, eliminating another piece of test equipment from the designer's bench. Additionally, the batteries provide some mass, which is needed to keep the board from falling off the bench.

Three banana jacks are mounted at the front of the board. These connect to the DUT ground and the low-impedance and high-impedance sides of the resistor. The resistor should be on the DUT and should preferably be built in to the pc board. There is negligible cost or mass penalty to one extra resistor in the assembly. Test points on either side of the resistor are extremely helpful for clipping in both the banana test leads and the scope probes. A three-pin SIP header is very useful for connecting all three signals at once.

When making plant measurements, the banana leads stay as they are, but the measurement BNC is unplugged, and the measurement test node is probed directly from the network analyzer to the node of interest on the DUT. Other useful user features on the board include a power-on LED and a low-battery LED. Both are very helpful at preventing measurement errors.

CLOSED LOOP OPEN LOOP CONTROL SYSTEM

Under Control System

When a number of elements are combined together to form a system to produce desired output then the system is referred as control system. As this system controls the output, it is so referred. Each element connected to the system has its own effect on the output.

Definition of Control System

A **control system** is a system of devices or set of devices, that manages, commands, directs or regulates the behavior of other device(s) or system(s) to achieve desire results. In other words the **definition of control system** can be rewritten as **A control system is a system, which controls other system**.

As the human civilization is being modernized day by day the demand of automation is increasing accordingly. Automation highly requires control of devices. In recent years, **control systems** plays main role in the development and advancement of modern technology and civilization. Practically every aspects of our day-to-day life is affected less or more by some control system. A bathroom toilet tank, a refrigerator, an air conditioner, a geezer, an automatic iron, an automobile all are control system. These systems are also used in industrial process for more output. We find control system in quality control of products, weapons system, transportation systems, power system, space technology, robotics and many more. The **principles of control theory** is applicable to engineering and non-engineering field both.

Feature of Control System

The main feature of control system is, there should be a clear mathematical relation between input and output of the system. When the relation between input and output of the system can be represented by a linear proportionality, the system is called linear control system. Again when the relation between input and output cannot be represented by single linear proportionality, rather the input and output are related by some non-linear relation, the system is referred as non-linear control system.

Requirement of Good Control System

Accuracy: Accuracy is the measurement tolerance of the instrument and defines the limits of the errors made when the instrument is used in normal operating conditions. Accuracy can be improved by using feedback elements. To increase accuracy of any control system error detector should be present in control system.

Sensitivity: The parameters of control system are always changing with change in surrounding conditions, internal disturbance or any other parameters. This change can be expressed in terms of sensitivity. Any control system should be insensitive to such parameters but sensitive to input signals only.

Noise: An undesired input signal is known as noise. A good control system should be able to reduce the noise effect for better performance.

Stability: It is an important characteristic of control system. For the bounded input signal, the output must be bounded and if input is zero then output must be zero then such a control system is said to be stable system.

Bandwidth: An operating frequency range decides the bandwidth of control system. Bandwidth should be large as possible for frequency response of good control system.

Speed: It is the time taken by control system to achieve its stable output. A good control system possesses high speed. The transient period for such system is very small.

Oscillation: A small numbers of oscillation or constant oscillation of output tend to system to be stable.

Types of Control Systems

There are various **types of control system** but all of them are created to control outputs. The system used for controlling the position, velocity, acceleration, temperature, pressure, voltage and current *etc.* are examples of control systems. Let us take an example of simple temperature controller of the room, to clear the concept. Suppose there is a simple heating element, which is heated up as long as the electric power supply is switched on.

As long as the power supply switch of the heater is on the temperature of the room rises and after achieving the desired temperature of the room, the power supply is switched off. Again due to ambient temperature, the room temperature falls and then manually the heater element is switched on to achieve the desired room temperature again. In this way one can manually control the room temperature at desired level.

This is an example of **manual control system**. This system can further be improved by using timer switching arrangement of the power supply where the supply to the heating element is switched on and off in a predetermined interval to achieve desired temperature level of the room. There is another improved way of controlling the temperature of the room. Here one sensor measures the difference between actual temperature and desired temperature. If there is any difference between them, the heating element functions to reduce the difference and when the difference becomes lower than a predetermined level, the heating elements stop functioning.

Both forms of the system are **automatic control system**. In former one the input of the system is entirely independent of the output of the system. Temperature of the room (output) increases as long as the power supply switch is kept on. That means heating element produces heat as long as the power supply is kept on and final room temperature does not have any control to the input power supply of the system. This system is referred as **open loop control system**. But in the later case, the heating elements of the system function, depending upon the difference between, actual temperature and desired temperature. This difference is called error of the system. This error signal is fed back to the system to control the input. As the input to output path and the error feedback path create a closed loop, this type of control system is referred as **closed loop control system**.

Hence, there are two main **types of control system**. They are as follow **Open loop control system Closed loop control system**

Open Loop Control System

A control system in which the control action is totally independent of output of the system then it is called **open loop control system**. Manual control system is also an open loop control system. Figure shows the block diagram of open loop control system in which process output is totally independent of controller action.

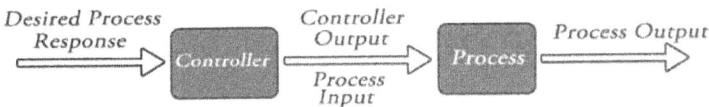

Practical Examples of Open Loop Control System

1. **Electric Hand Drier** – Hot air (output) comes out as long as you keep your hand under the machine, irrespective of how much your hand is dried.
2. **Automatic Washing Machine** – This machine runs according to the pre-set time irrespective of washing is completed or not.
3. **Bread Toaster** - This machine runs as per adjusted time irrespective of toasting is completed or not.
4. **Automatic Tea/Coffee Maker** – These machines also function for pre adjusted time only.

5. **Timer Based Clothes Drier** – This machine dries wet clothes for pre – adjusted time, it does not matter how much the clothes are dried.

6. **Light Switch** – lamps glow whenever light switch is on irrespective of light is required or not.

7. **Volume on Stereo System** – Volume is adjusted manually irrespective of output volume level.

Advantages of Open Loop Control System

1. Simple in construction and design.
2. Economical.
3. Easy to maintain.
4. Generally stable.
5. Convenient to use as output is difficult to measure.

Disadvantages of Open Loop Control System

1. They are inaccurate.
2. They are unreliable.
3. Any change in output cannot be corrected automatically.

Closed Loop Control System

Control system in which the output has an effect on the input quantity in such a manner that the input quantity will adjust itself based on the output generated is called **closed loop control system**. Open loop control system can be converted in to closed loop control system by providing a feedback. This feedback automatically makes the suitable changes in the output due to external disturbance. In this way closed loop control system is called automatic control system. Figure below shows the block diagram of closed loop control system in which feedback is taken from output and fed in to input.

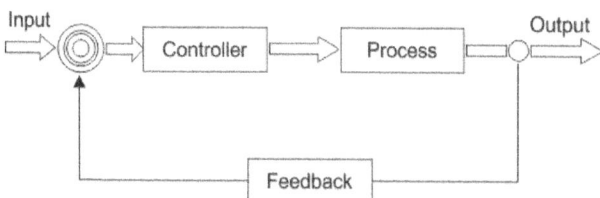

Practical Examples of Closed Loop Control System

1. **Automatic Electric Iron** – Heating elements are controlled by output temperature of the iron.

2. **Servo Voltage Stabilizer** – Voltage controller operates depending upon output voltage of the system.

3. **Water Level Controller–** Input water is controlled by water level of the reservoir.

4. **Missile Launched & Auto Tracked by Radar –** The direction of missile is controlled by comparing the target and position of the missile.

5. **An Air Conditioner –** An air conditioner functions depending upon the temperature of the room.

6. **Cooling System in Car –** It operates depending upon the temperature which it controls.

Advantages of Closed Loop Control System

1. Closed loop control systems are more accurate even in the presence of non-linearity.

2. Highly accurate as any error arising is corrected due to presence of feedback signal.

3. Bandwidth range is large.

4. Facilitates automation.

5. The sensitivity of system may be made small to make system more stable.

6. This system is less affected by noise.

Disadvantages of Closed Loop Control System

1. They are costlier.

2. They are complicated to design.

3. Required more maintenance.

4. Feedback leads to oscillatory response.

5. Overall gain is reduced due to presence of feedback.

6. Stability is the major problem and more care is needed to design a stable closed loop system.

Table: Comparison of Closed Loop And Open Loop Control System

Sr. No.	Open loop control system	Closed loop control system
1	The feedback element is absent.	The feedback element is always present.
2	An error detector is not present.	An error detector is always present.
3	It is stable one.	It may become unstable.
4	Easy to construct.	Complicated construction.
5	It is an economical.	It is costly.
6	Having small bandwidth.	Having large bandwidth.
7	It is inaccurate.	It is accurate.
8	Less maintenance.	More maintenance.
9	It is unreliable.	It is reliable.
10	Examples: Hand drier, tea maker	Examples: Servo voltage stabilizer, perspiration

Feedback Loop of Control System

A feedback is a common and powerful tool when designing a control system. Feedback loop is the tool which take the system output into consideration and enables the system to adjust its performance to meet a desired result of system.

In any control system, output is affected due to change in environmental condition or any kind of disturbance. So one signal is taken from output and is fed back to the input. This signal is compared with reference input and then error signal is generated. This error signal is applied to controller and output is corrected. Such a system is called feedback system. Figure below shows the block diagram of feedback system.

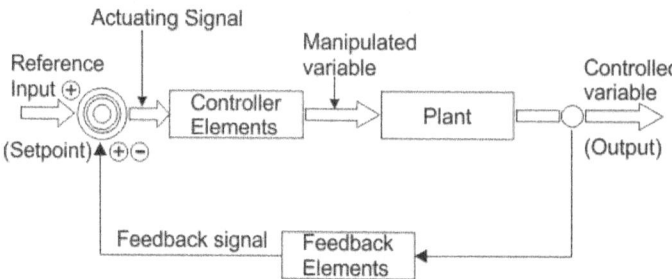

When feedback signal is positive then system called positive feedback system. For positive feedback system, the error signal is the addition of reference input signal and feedback signal. When feedback signal is negative then system is called negative feedback system. For negative feedback system, the error signal is given by difference of reference input signal and feedback signal.

Effect of Feedback

Refer figure beside, which represents feedback system where R = Input signal E = Error signal G = forward path gain H = Feedback C = Output signal B = Feedback signal

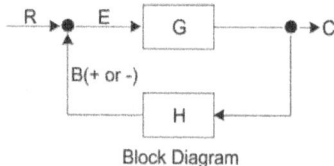

Block Diagram

1. Error between system input and system output is reduced.
2. System gain is reduced by a factor $1/(1 \pm GH)$.
3. Improvement in sensitivity.
4. Stability may be affected.
5. Improve the speed of response.

FEEDBACK LOOP ANALYSIS OF POWER SUPPLY CONTROL LOOPS

Every power supply has a feed-back loop that monitors the output voltage or current and keeps the device's output level constant despite changes in the load. This means that the power device conducts longer if the output voltage is too low. Most switched mode power supplies use pulse width (PWM) or frequency modulation (FM) in their control loops. Analysis of the loop dynamics requires the ability to demodulate these signals. Power Analysis software, available in Teledyne LeCroy oscilloscopes, includes easy-to-use modulation analysis capabilities.

Modulation analysis functions produce a time domain display that represents the modulated parameter in a time vs. time graphical plot. They are convenient tools for intuitively viewing the time domain response of the entire control loop, including any time constants added by the pulse width modulator. Modulation analysis can be performed for duty cycle, period, frequency, or pulse width.

An example is shown in Figure, where the response to a step load change of a PWM-based control loop is shown.

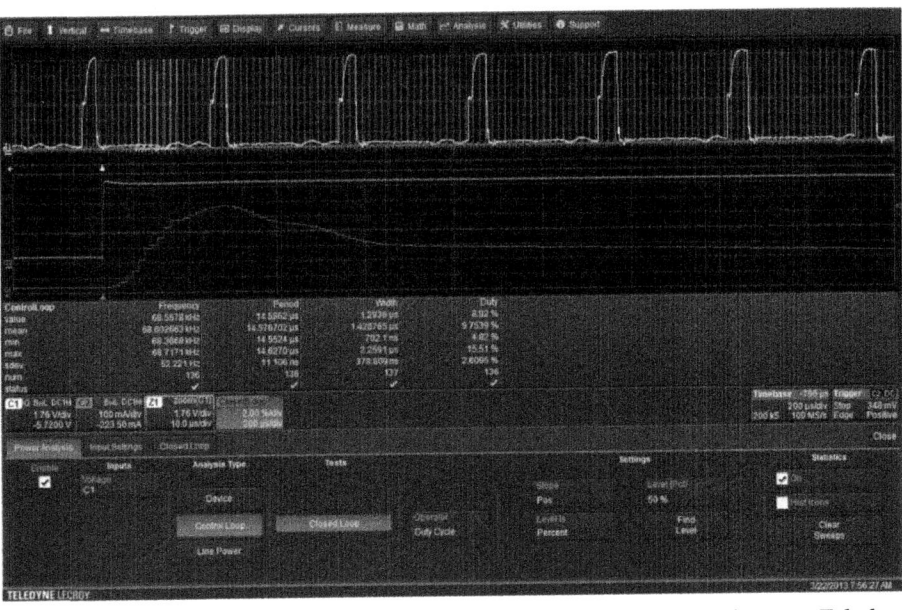

Fig. : Measuring the step load response in a switched mode power supply using Teledyne LeCroy\'s Power Analysis option software

The upper trace, C1, is the gate-to-source drive signal to a MOSFET. This PWM signal is demodulated using a track function of width shown in the lower Control Loop trace. The Control Loop function displays the duty cycle of the gate drive signal as a function of time, which is time synchronous with the source waveform. The user can use the zoom features and see the width of each individual cycle and the corresponding value of the track plot so each point in the track function can be related to the source waveform.

It is easy to see that the control loop initially overshoots and then recovers in about 800 μs. The time scale of this acquisition is 200 μs per division and the vertical scaling of the track of width is 2% per division. The pulse duty cycle before the load change is approximately 4.8%. After the change, it increases to 15% and then quickly recovers to 9%.

Note that the use of a moderate acquisition memory length of 200 kilo Samples allows the measured waveforms to be digitized at 100 MS/s for a time resolution of 10 ns. This particular oscilloscope offers a maximum of 250MS memory length. Since the switching frequency of this supply is only 68 kHz, the 100 MS/s sample rate is more than adequate to provide ample time resolution in the measurement.

A related study is shown in Figure. In this example, Power Analysis software running on an HDO 6000 oscilloscope acquires a 20 ms record including every gate drive pulse from the time the power supply is turned on until it reaches steady state.

Fig. : A control loop study showing the startup of the power supply.

The modulation analysis display shows the pulse width of every cycle of the gate drive signal as it occurs. The soft start circuit's performance is readily observed. The minimum and maximum parameters read the range of pulse width variations as 220 ns to 5.08 μs.

The Power Analysis option simplifies power analysis by automating the setup of even these relatively complex functions. Modulation analysis can be used to characterize power supply stability under load changes, line changes, soft-starts, dropouts, hot swap, and short circuits. It allows us to see, on a cycle-by-cycle basis, the behavior of the control loop.

ANALYZING AND IMPROVING CONTROL LOOP PERFORMANCE

Power plants—especially coal-fired boilers—pose a set of process control problems unparalleled in other industries. Rapid load ramps, continuous unit demand changes, variable fuel quality and process interactions are only a few of the factors challenging the performance of boiler control loops. Attempts to address poor performance are often limited to the application of trial-and-error tuning methods and frequently fall short of the objectives.

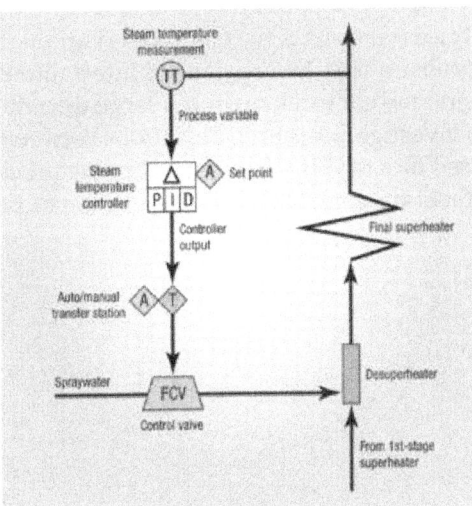

Fig. : Simple Steam Temperature Control Loop.

Although some control problems can potentially be resolved through controller tuning, poor control loop performance can have several other causes requiring different solutions. Once the origin of the problem has been correctly identified, the solution is frequently obvious. The origin of the problem can be pinpointed through a simple but systematic analysis of the control problem.

Poor control loop performance often makes itself evident as excessively large deviations between the process variable and its set point. An operator may notice the deviations on his process trend displays or process alarm system and place the controller into manual mode in an attempt to stabilize the control loop. As a result, poor control loop performance may be indicated by excessive deviations between process variable and set point or controllers being in manual mode.

Controllers in Manual

Some plants have 30 percent or more of their control loops in manual control mode. Being in manual mode does not necessarily mean the control loop performs poorly in automatic control mode. Many control loops are associated with redundant equipment, or certain operating modes, and can justifiably be in manual mode as a result. However, if a control loop is supposed to be in automatic control mode, but it is in manual mode, an investigation is required. It is important to talk to the operator to find out why the loop is in manual. Historical process trends of times when the loop was in automatic control can be reviewed to gain more insight into the problem, or the loop can be placed into automatic control mode to observe how it responds.

Deviations from Set Point

If a controller is in automatic control mode and its process variable always remains acceptably close to its set point, there should be no need for concern. However, excessively large deviations between process variable and set point indicate poor control. The problem may be constant or intermittent and could originate from within or outside the control loop, but if large deviations occur, it provides grounds for further investigation. Large deviations between process variable and set point can be caused by a rapidly changing set point, process disturbances, loop nonlinearities, interactions, control element saturation or poor controller tuning.

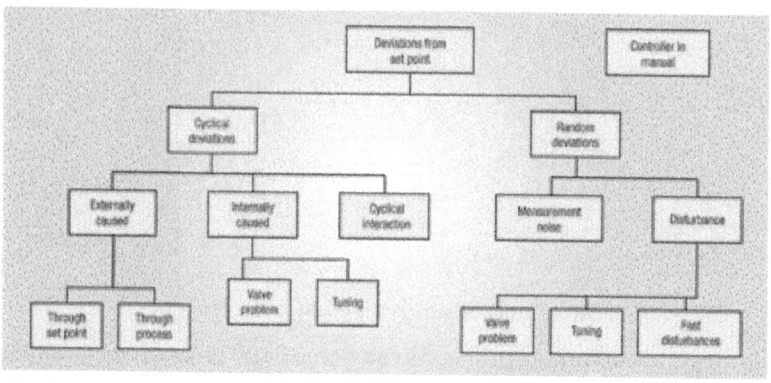

Fig. : Fault Analysis Tree for Systematically Determining the Cause of Poor Control.

The first step in analyzing the control problem would be to look at the shape of the deviations on a trend plot: whether they are cyclical (oscillating) or random. This will determine the path for further analysis. Although sophisticated frequency analysis tools can be used to determine if deviations are cyclical or random, it can also be done quite simply by looking at historical time trends of the process variable. Once it has been determined if the deviations from set point are cyclical or not, the next level of analysis can be conducted.

Cyclical Deviations

Cyclical deviations can appear as sine, saw tooth, or square wave patterns. A cyclical deviation between set point and process variable can originate from within the control loop or it could be caused by external factors. It could also be a result of a cyclical interaction between two or more control loops. To narrow down the cause of the oscillation, the controller should be placed into manual mode to see if the oscillation would stop. If the oscillation persists when the controller is in manual mode, it originates from outside the loop.

Externally-caused Oscillations

An oscillation with its origin outside the control loop can influence the control loop through its set point or through the process.

Oscillations entering the loop through the set point are easy to find; simply look at where the loop's set point is driven from. For example, if a feedwater flow control loop oscillates because its set point is oscillating, the problem lies with the drum level controller. The fault analysis should then be applied to the drum level controller.

Oscillating Process

One oscillating loop can cause several other loops on the same plant to oscillate with it. For example, if the steam pressure controller on a boiler oscillates, several other loops including the steam temperature will likely oscillate too. The steam temperature will keep on oscillating, even if the steam temperature controller is placed into manual control mode. The loops will all oscillate in harmony with the same period of oscillation.

Historical trends or process analysis software can be used to identify all the loops oscillating with the same period. The problem loop can then be isolated through knowledge of the boiler systems and their interactions, or by placing likely culprit loops in manual one at a time. If the loop driving the oscillations is placed into manual control mode, the oscillations will cease on all loops. This loop should then be analyzed further.

Note that this scenario is different from a cyclical interaction in which two or more control loops interact directly with each other in a cyclical fashion. In the case of a cyclical interaction, any of the participating control loops placed into manual will cause all loops to stop oscillating. This will be discussed later.

Internally-Caused Oscillations

Oscillations generated by a control loop itself can be caused by faulty final control element (for example, control valve or damper) or by tuning. Generally, if the oscillation is caused by poor tuning, the process variable will oscillate with a reasonably smooth sine-wave pattern. If the oscillation is caused by final control element problems, the trends are more likely to be shaped like a square wave or saw tooth wave. However, this is a guideline and not a definitive test. If the control loop drives a final control element, the performance of the latter should be checked first before attempting to tune the controller. This is especially true if the control loop used to work properly and is now oscillating without any changes to the controller settings.

The most common equipment-based causes of oscillations are control valve (or damper) related. Note that the discussions below sometimes mention only control valves for the sake of brevity. However, dampers can cause the same problems with the same symptoms as control valves. So where only control valves are mentioned, the same arguments will also apply to dampers.

Control Valve Problems

A common problem found in final control elements is stiction. This is short for Static Friction, and means that the valve internals are sticky. If the stem of a valve with stiction comes to rest, it tends to stick in that position. Additional force is then required to overcome the stiction.

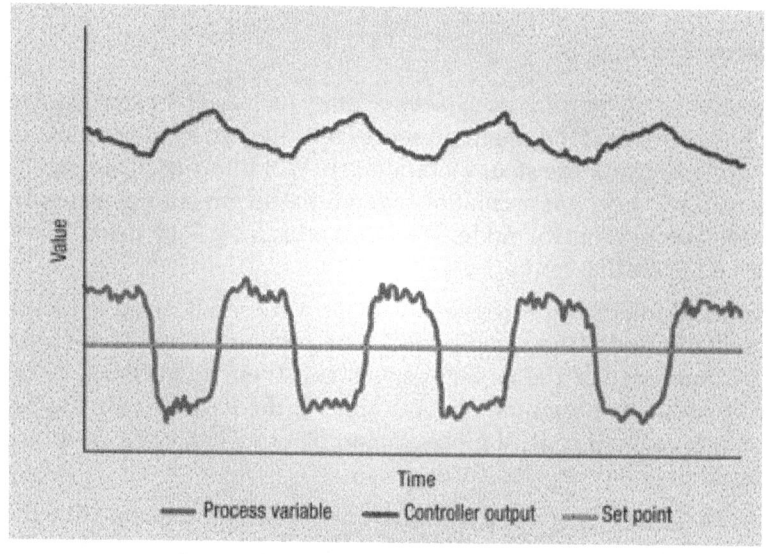

Fig. : A Flow Loop with a Stick-Sup Cycle.

A controller in automatic control mode will continue to change its output in an attempt to get the process variable to its set point. While the valve is sticking, the process remains deviated from set point but additional pressure builds up in the valve actuator. If enough pressure has been built up to overcome the static friction, the valve breaks free and travels to the new controller output which is now far beyond its original value. This causes the process to overshoot its set point. Then the valve sticks at the new position, the controller output reverses its direction of travel and the whole process repeats in the opposite direction. This causes an oscillation, called a stick-slip cycle. If loop oscillations are caused by stiction, the controller output's cycle often resembles a saw-tooth wave, while the process variable may look like a square wave or an irregular sine wave.

Stiction might be caused by an over-tight valve stem seal, by sticky valve internals, by an undersized actuator, or a faulty positioner. Stiction can be detected by placing the controller in manual mode and making small changes (0.5 percent is recommended) in controller output and monitoring the process variable for a resulting change. If the control valve seems to accumulate a few of the controller output changes before the process variable shows movement, it has stiction.

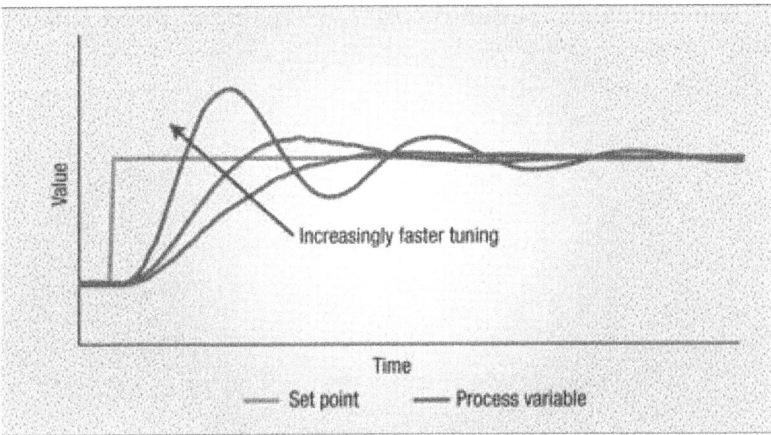

Fig. : The Settling Time of a Control Loop has a Minimum Limit Beyond which the Loop Begins to Oscillate.

Because of the widespread adoption of positioners for accurately positioning control valves and dampers, one problem that is more common now than a decade ago, is that of positioner overshoot. Positioners are fast feedback controllers mounted on the final control element to measure the valve stem or damper vane position and manipulate the actuator until the desired valve position is achieved. Most positioners can be tuned. Some are tuned too aggressively for the valve or damper they are controlling. This causes the device to overshoot its target position after a change in controller output. The positioner may cause the valve or damper position to hunt around or even oscillate. Sometimes the positioner is simply defective and causes the device to overshoot. If the controller on a fast-responding loop like a flow control loop is tuned aggressively, the combination with positioner overshoot can cause severe oscillations in the control loop. Positioner overshoot can be detected on fast-responding loops by placing the controller in manual and changing the controller output by 2 to 5 percent.

If a level controller drives a valve directly (*i.e.* no cascade control), and the valve has dead band, the loop will continuously oscillate. Valve dead band will be discussed later. A level control loop will also oscillate if the controller has an internal dead band around the set point. This is sometimes done to prevent the controller from reacting to process noise, but it should not be done on level loops because of the continuous oscillation it causes.

Tuning

A loop that is tuned too aggressively (overly fast response) can quickly develop oscillations. Step tests should be done on the process to determine the dominant process characteristics: process gain, dead time and time constant. A step test is done by placing the controller into manual mode and changing its output by a few percent (between 2 and 5 percent is normally sufficient). Three or more step tests should be done to compare the results, throw out outliers and use the average.

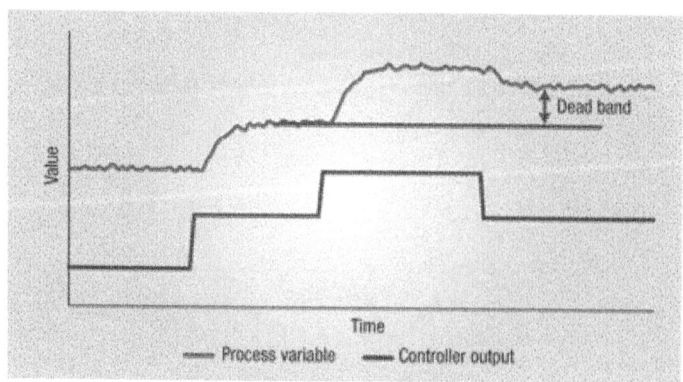

Fig. : Test Revealing The Presence of Dead Band in a Control Valve.

Proven, broad-spectrum tuning rules like the Cohen-Coon or Lambda tuning rules should be used to calculate new controller settings. However, the Cohen-Coon tuning rules are too aggressive in their original form, and it is recommended to use only half of the calculated controller gain. Best practices prescribe using tuning software for analyzing step-test data and calculating new controller settings.

Cyclical Interaction

Interaction between loops with similar dynamics can cause the two loops to "fight" each other. One example of this is the main steam temperature controller interacting with the unit load controller or the steam pressure controller. If the steam temperature is too high, the steam temperature controller injects more spraywater into the desuperheaters. This increases steam production that will either increase unit load or steam pressure, depending on the front-end control mode. In any event, the fuel firing rate will be reduced, the steam temperature cools, and the whole effect reverses and cycles.

Cyclical interaction is aggravated by aggressive tuning. Most boiler control loops are traditionally tuned aggressively to obtain a fast response during load changes and boiler upsets. Because of the highly interactive nature of boiler sub-systems, many boilers are prone to cyclical interactions.

To solve problems with cyclical interactions, control loops have to be tuned less aggressively. Using the Lambda method for tuning controllers results in very stable control loops. One can think of highly interactive control loops as a tub filled with water. If you drop a stone in the tub, lots of waves result that take a while to stabilize. Using the Lambda tuning method is like replacing the water with oil. Now if you drop the stone into the tub, the oil just absorbs the disturbance and no persistent wave action results. The pulp and paper industry is also plagued with highly interactive processes, and it has had great success using the Lambda tuning method.

Random Deviations

In contrast to oscillations that are periodic, poor control can also make itself evident in large but random deviations between the process variable and set point. These could be measurement noise, process disturbances or rapid set point changes. To understand how a control loop is capable of handling disturbances, we need to look at the speed of response of a control loop.

There are several measurements for loop response; settling time will be used here. Settling time can be defined as the duration of time during which a deviation between set point and process variable is more than 5 percent of the size of the deviation. The settling time of a control loop cannot be infinitely short. If a control loop is tuned sluggishly, it will have a long settling time. If the tuning is improved, the settling time will be reduced, but only up to a point. If the tuning is made any faster, the loop will become cyclical and the settling time will increase.

The minimum settling time of a control loop is determined mostly by the amount of dead time in the process. For a flow loop, the settling time is about three times the dead time, for a temperature loop it's between three and four dead times, and for a level loop it is about four dead times.

Measurement Noise

Measurement noise is deviations from set point that change direction very rapidly. The rate at which this happens is so much shorter than the loop settling time that it is impossible for the controller to eliminate noise or even reduce its amplitude. A controller responding aggressively to noise will likely increase the average deviation size. The amplitude of noise can be reduced through filtering the process variable with a first-order lag filter. It is important to note that a filter increases the apparent dead time of a loop and therefore increases its settling time. Filtering should be applied only when needed, and then as little of it as possible.

Disturbances

A process disturbance can push the process variable away from its set point. Disturbances are often the nemesis of good loop performance. As described above, feedback control is limited in how fast it can eliminate the effects of a disturbance and bring the process back to set point. If the disturbance occurs much slower than the settling time of a control loop, feedback control should be able to significantly reduce its amplitude. If not, it may be a problem with the final control element or the tuning of the controller. One should first check for final control element problems before tuning the controller.

Dead band (sometimes called hysteresis), reduces the effectiveness with which a controller can counteract disturbances. Every time the process variable undergoes a disturbance in a different direction from the previous disturbance, the controller output has to traverse the entire dead band before the valve or damper begins moving. Dead band can be detected very reliably with a simple process test consisting of two controller output steps in one direction and one step in the opposite direction with the controller in manual mode. The second and last steps should be the same size. If the process variable does not reach the same level after the first and third steps, it indicates the presence of dead band. Dead band is a mechanical problem and cannot be addressed with tuning.

A control loop may also appear to have sluggish response if the controller output becomes saturated at its upper or lower limit. If the controller output is constrained by a rate-of-change limiter, it also may cause sluggish response regardless of how well the controller is tuned. Alarms can warn of these conditions or historical time-trends of the controller output and process variable can be reviewed to find their presence.

Tuning

Once the final control element has a clean bill of health, the controller tuning should be reviewed to see if it is perhaps sluggish tuning that reduces the controller's effectiveness in counteracting disturbances. Controller tuning advice given earlier applies here too. Note that the Lambda tuning method results in stable control loops, but often cause a sluggish response to disturbances, especially on slow temperature loops. Cohen-Coon tuning provides faster disturbance rejection.

Although correct tuning methods can go a long way in minimizing the effects of disturbances, disturbances sometimes happen so rapidly that feedback control alone is unable to reduce their effects to reasonable levels. Realize that feedback control has a limit to the speed of response. Once this limit has been reached, other solutions must be sought to obtain further improvement in performance. It is sad to hear of personnel spending days and even weeks tweaking a control loop that is already at the limit of its performance capability.

Rapidly-Changing Disturbances

If a disturbance occurs faster than the control loop can respond, there is very little the controller can do to reduce its amplitude. In cases like this the feedback controller can be greatly augmented with cascade and feedforward control. Although the standard design of boiler controls includes several cascade and feedforward controls, they have sometimes not been tuned for optimal response.

A good example of the application of cascade control is the main steam temperature controller cascaded with the desuperheater outlet temperature controller. The latter virtually eliminates disturbances coming from changes in spraywater pressure and temperature upsets coming from the first-stage superheater. There are many more examples in standard boiler control configurations, and often a few additional opportunities for improving boiler controls.

Applications of feedforward control include the feedforward from the steam flow measurement to the feedwater flow controller, forming part of the three element drum level control. Another example is the feedforward between the total air flow and the ID Fans.

Some control problems seem to come and go with time—intermittent problems. These problems are more difficult to track down and solve, but it helps to know what causes to look for.

Nonlinear valve Characteristic

Many control valves and most dampers have a nonlinear installed characteristic. This means that the flow characteristic of the device changes depending on how much open it is. If tuning is done with the valve or damper at the one end of its travel, the settings might not work at the other end and could cause oscillations or sluggish behavior. If this is the case, a function generator (X-Y curve) can be placed in the path of the controller output to cancel out the control valve or damper nonlinearity. Boiler control designs often include function generators, but in some cases these have never been calibrated and still contain the original linear curve.

Nonlinear Process

Many boiler subsystems react differently based on unit load, mills in service, soot levels, *etc.* In many cases the differences in process characteristics are large enough to affect control loop performance. For example, the cooling effect of desuperheater spray flow is much less at high steam flow rates compared to low steam flow rates. These changes in process characteristics often require different tuning settings for optimal control at various operating conditions. However, this

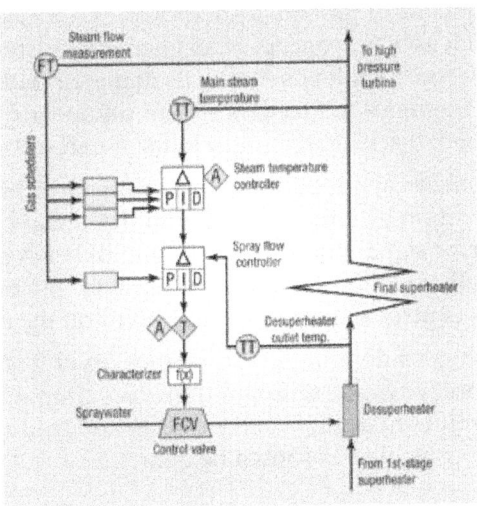

Fig. : Advanced Steam Temperature Control.

is seldom implemented, leaving the control loop with poor response for most of its operating range. On systems with varying process characteristics, controller tuning should be altered automatically based on the operating conditions. This is accomplished quite effectively by implementing gain scheduling. Gain scheduling, as seen in Figure, uses the operating condition (like steam flow rate) as an input to one or more function generators to dynamically adjust the controller gain, and sometimes also the integral time, and derivative time if used.

Obtaining robust and properly performing boiler controls can be very challenging. Control loops can perform sub-optimally due to a variety of reasons and controller tuning alone is in many cases not the ultimate remedy for poor control performance. Through a simple but systematic analysis of the control problem, the root cause of poor control can be established and the problem can be resolved or at least minimized in the most effective way.

CLOSED-LOOP SYSTEMS

In the previous tutorial we saw that systems in which the output quantity has no effect upon the input to the control process are called open-loop control systems, and that open-loop systems are just that, open ended non-feedback systems. But the goal of any electrical or electronic control system is to measure, monitor, and control a process.

One way in which we can accurately Control the Process is by monitoring its output and "feeding" some of it back to compare the actual output with the desired output so as to reduce the error and if disturbed, bring the output of the system back to the original or desired response. The measure of the output is called the "feedback signal" and the type of control system which uses feedback signals to control itself is called a Close-loop System.

A Closed-loop Control System, also known as a *feedback control system* is a control system which uses the concept of an open loop system as its forward path but has one or more feedback loops (hence its name) or paths between its output and its input. The reference to "feedback", simply means that some portion of the output is returned "back" to the input to form part of the systems excitation.

Closed-loop systems are designed to automatically achieve and maintain the desired output condition by comparing it with the actual condition. It does this by generating an error signal which is the difference between the output and the reference input. In other words, a "closed-loop system" is a fully automatic control system in which its control action being dependent on the output in some way.

So for example, consider our electric clothes dryer from the previous open-loop tutorial. Suppose we used a sensor or transducer (input device) to continually monitor the temperature or dryness of the clothes and feed a signal relating to the dryness back to the controller as shown below.

Closed-loop Control

This sensor would monitor the actual dryness of the clothes and compare it with (or subtract it from) the input reference. The error signal (error = required dryness – actual dryness) is amplified by the controller, and the controller output makes the necessary correction to the heating system to reduce any error. For example if the clothes are too wet the controller may increase the temperature or drying time. Likewise, if the clothes are nearly dry it may reduce the temperature or stop the process so as not to overheat or burn the clothes, *etc.*

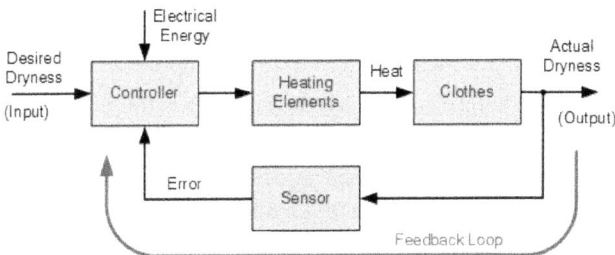

Then the closed-loop configuration is characterised by the feedback signal, derived from the sensor in our clothes drying system. The magnitude and polarity of the resulting error signal, would be directly related to the difference between the required dryness and actual dryness of the clothes.

Also, because a closed-loop system has some knowledge of the output condition, (via the sensor) it is better equipped to handle any system disturbances or changes in the conditions which may reduce its ability to complete the desired task.

For example, as before, the dryer door opens and heat is lost. This time the deviation in temperature is detected by the feedback sensor and the controller self-corrects the error to maintain a constant temperature within the limits of the preset value. Or possibly stops the process and activates an alarm to inform the operator.

As we can see, in a closed-loop control system the error signal, which is the difference between the input signal and the feedback signal (which may be the output signal itself or a function of the output signal), is fed to the controller so as to reduce the systems error and bring the output of the system back to a desired value. In our case the dryness of the clothes. Clearly, when the error is zero the clothes are dry.

The term Closed-loop control always implies the use of a feedback control action in order to reduce any errors within the system, and its "feedback" which distinguishes the main differences between an open-loop and a closed-loop system.

The accuracy of the output thus depends on the feedback path, which in general can be made very accurate and within electronic control systems and circuits, feedback control is more commonly used than open-loop or feed forward control.

Closed-loop systems have many advantages over open-loop systems. The primary advantage of a closed-loop feedback control system is its ability to reduce a system's sensitivity to external disturbances, for example opening of the dryer door, giving the system a more robust control as any changes in the feedback signal will result in compensation by the controller.

Then we can define the main characteristics of Closed-loop Control as being:

- To reduce errors by automatically adjusting the systems input.
- To improve stability of an unstable system.
- To increase or reduce the systems sensitivity.
- To enhance robustness against external disturbances to the process.
- To produce a reliable and repeatable performance.

Whilst a good closed-loop system can have many advantages over an open-loop control system, its main disadvantage is that in order to provide the required amount of control, a closed-loop system must be more complex by having one or more feedback paths. Also, if the gain of the controller is too sensitive to changes in its input commands or signals it can become unstable and start to oscillate as the controller tries to over-correct itself, and eventually something would break. So we need to "tell" the system how we want it to behave within some pre-defined limits.

Closed-loop Summing Points

For a closed-loop feedback system to regulate any control signal, it must first determine the error between the actual output and the desired output. This is achieved using a summing point, also referred to as a comparison element, between the feedback loop and the systems input. These summing points compare a systems set point to the actual value and produce a positive or negative error signal which the controller responds too. where: Error = Set point – Actual

The symbol used to represent a summing point in closed-loop systems block-diagram is that of a circle with two crossed lines as shown. The summing point can either add signals together in which a Plus (+) symbol is used showing the device to be a "summer" (used for positive feedback), or it can subtract signals from each other in which case a Minus (–) symbol is used showing that the device is a "comparator" (used for negative feedback) as shown.

Summing Point Types

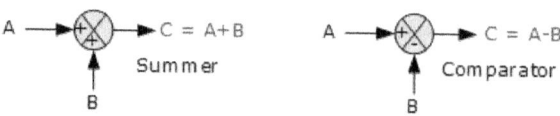

Note that summing points can have more than one signal as inputs either adding or subtracting but only one output which is the algebraic sum of the inputs. Also the arrows indicate the direction of the signals. Summing points can be cascaded together to allow for more input variables to be summed at a given point.

Closed-loop System Transfer Function

The Transfer Function of any electrical or electronic control system is the mathematical relationship between the systems input and its output, and hence describes the behaviour of the system. Note also that the ratio of the output of a particular device to its input represents its gain. Then we can correctly say that the output is always the transfer function of the system times the input. Consider the closed-loop system below.

Typical Closed-loop System Representation

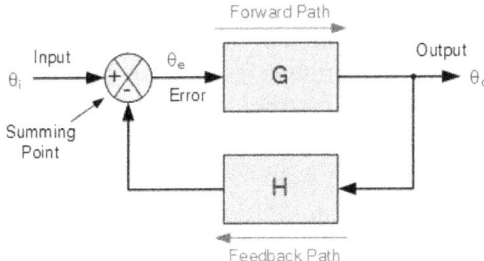

Where: block G represents the open-loop gains of the controller or system and is the forward path, and block H represents the gain of the sensor, transducer or measurement system in the feedback path.

To find the transfer function of the closed-loop system above, we must first calculate the output signal θo in terms of the input signal θi. To do so, we can easily write the equations of the given block-diagram as follows.

The output from the system is equal to: Output = G x Error

Note that the error signal, θe is also the input to the feed-forward block: G

The output from the summing point is equal to: Error = Input - H x Output

If H = 1 (unity feedback) then:

The output from the summing point will be: Error (θe) = Input - Output

Eliminating the error term, then:

The output is equal to: Output = G x (Input - H x Output)

Therefore: G x Input = Output + G x H x Output

Rearranging the above gives us the closed-loop transfer function of:

$$\frac{\text{Output}}{\text{Input}} = \frac{\theta_o}{\theta_i} = \frac{G}{1 + GH}$$

The above equation for the transfer function of a closed-loop system shows a Plus (+) sign in the denominator representing negative feedback. With a positive feedback system, the denominator will have a Minus (−) sign and the equation becomes: 1 - GH.

We can see that when H = 1 (unity feedback) and G is very large, the transfer function approaches unity as:

$$\frac{\text{Output}}{\text{Input}} \rightarrow 1$$

Also, as the systems steady state gain G decreases, the expression of: $G/(1 + G)$ decreases much more slowly. In other words, the system is fairly insensitive to variations in the systems gain represented by G, and which is one of the main advantages of a closed-loop system.

Multi-loop Closed-loop System

Whilst our example above is of a single input, single output closed-loop system, the basic transfer function still applies to more complex multi-loop systems. Most practical feedback circuits have some form of multiple loop control, and for a multi-loop configuration the transfer function between a controlled and a manipulated variable depends on whether the other feedback control loops are open or closed.

Consider the multi-loop system below.

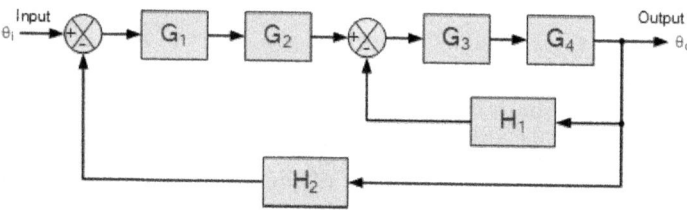

Any cascaded blocks such as G_1 and G_2 can be reduced, as well as the transfer function of the inner loop as shown.

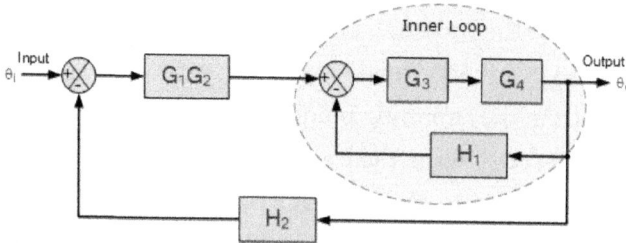

After further reduction of the blocks we end up with a final block diagram which resembles that of the previous single-loop closed-loop system.

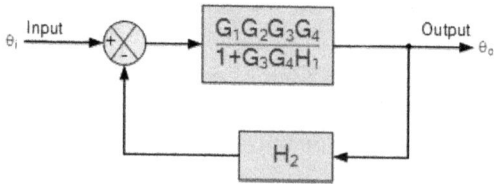

And the transfer function of this multi-loop system becomes:

$$\frac{Output}{Input} = \frac{\theta_o}{\theta_i} = \frac{G_1 G_2 G_3 G_4}{1 + G_3 G_4 H_1 + G_1 G_2 G_3 G_4 H_2}$$

Then we can see that even complex multi-block or multi-loop block diagrams can be reduced to give one single block diagram with one common system transfer function.

Closed-loop Motor Control

So how can we use Closed-loop Systems in Electronics. Well consider our DC motor controller from the previous open-loop tutorial. If we connected a speed measuring transducer, such as a tachometer to the shaft of the DC motor, we could detect its speed and send a signal proportional to the motor speed back to the amplifier. A tachometer, also known as a tacho-generator is simply a permanent-magnet DC generator which gives a DC output voltage proportional to the speed of the motor.

Then the position of the potentiometers slider represents the input, θ_i which is amplified by the amplifier (controller) to drive the DC motor at a set speed N representing the output, θ_o of the system, and the tachometer T would be the closed-loop back to the controller. The difference between the input voltage setting and the feedback voltage level gives the error signal as shown.

Closed-loop Motor Control

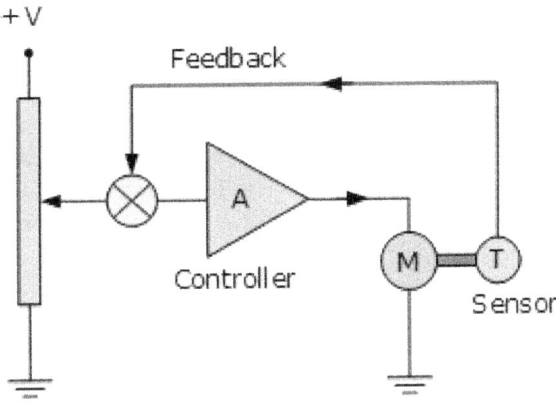

Any external disturbances to the closed-loop motor control system such as the motors load increasing would create a difference in the actual motor speed and the potentiometer input set point.

This difference would produce an error signal which the controller would automatically respond too adjusting the motors speed. Then the controller works to minimize the error signal, with zero error indicating actual speed which equals set point.

Electronically, we could implement such a simple closed-loop tachometer-feedback motor control circuit using an operational amplifier (op-amp) for the controller as shown.

Closed-loop Motor Controller Circuit

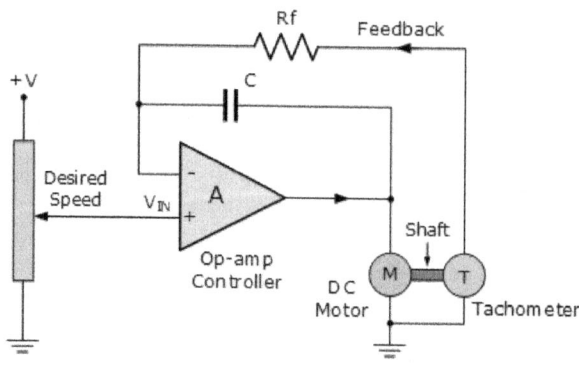

This simple closed-loop motor controller can be represented as a block diagram as shown.

Block Diagram for the Feedback Controller

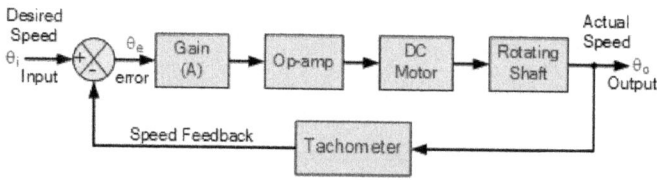

A closed-loop motor controller is a common means of maintaining a desired motor speed under varying load conditions by changing the average voltage applied to the input from the controller. The tachometer could be replaced by an optical encoder or Hall-effect type positional or rotary sensor.

Closed-loop Systems Summary

We have seen that an electronic control system with one or more feedback paths is called a Closed-loop System. Closed-loop control systems are also called "feedback control systems" are very common in process control and electronic control systems. Feedback systems have part of their output signal "fed back" to the input for comparison with the desired set point condition. The type of feedback signal can result either in positive feedback or negative feedback.

In a closed-loop system, a controller is used to compare the output of a system with the required condition and convert the error into a control action designed to reduce the error and bring the output of the system back to the desired response. Then closed-loop control systems use feedback to determine the actual input to the system and can have more than one feedback loop.

Closed-loop control systems have many advantages over open-loop systems. One advantage is the fact that the use of feedback makes the system response relatively insensitive to external disturbances and internal variations in system parameters such as temperature. It is thus possible to use relatively inaccurate and inexpensive components to obtain the accurate control of a given process or plant.

However, system stability can be a major problem especially in badly designed closed-loop systems as they may try to over-correct any errors which could cause the system to loss control and oscillate.

CLOSED-LOOP CONTROL

Many real-time embedded systems make control decisions. These decisions are usually made by software and based on feedback from the hardware under its control (termed the plant). Such feedback commonly takes the form of an analog sensor that can be read via an A/D converter. A sample from the sensor may represent position, voltage, temperature, or any other appropriate parameter. Each sample provides the software with additional information upon which to base its control decisions.

Closing the Loop

Systems that utilize feedback are called closed-loop control systems. The feedback is used to make decisions about changes to the control signal that drives the plant. An open-loop control system doesn't have or doesn't use feedback.

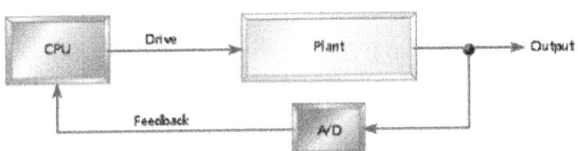

Fig. : A closed-loop control system.

A basic closed-loop control system is shown in Figure. This figure can describe a variety of control systems, including those driving elevators, thermostats, and cruise control.

Closed-loop control systems typically operate at a fixed frequency. The frequency of changes to the drive signal is usually the same as the sampling rate, and certainly not any faster. After reading each new sample from the sensor, the software reacts to the plant's changed state by recalculating and adjusting the drive signal. The plant responds to this change, another sample is taken, and the cycle repeats. Eventually, the plant should reach the desired state and the software will cease making changes.

If feedback indicates that the temperature in your home is below your desired setpoint, the thermostat will turn the heater on until the room is at least that temperature. Similarly, if your car is going too quickly, the cruise control system can temporarily reduce the amount of fuel fed to the engine.

Bang Bang

How much should the software increase or decrease the drive signal? One option is to just set the drive signal to its minimum value when you want the plant to decrease its activity and to its maximum value when you want the plant to increase its activity. This strategy is called on-off control, and it is how many thermostats work.

On-off control doesn't work well in all systems. If the thermostat waits until the desired temperature is achieved to turn off the heater, the temperature may overshoot. The same amount of overshoot and ripple probably isn't acceptable in an elevator.

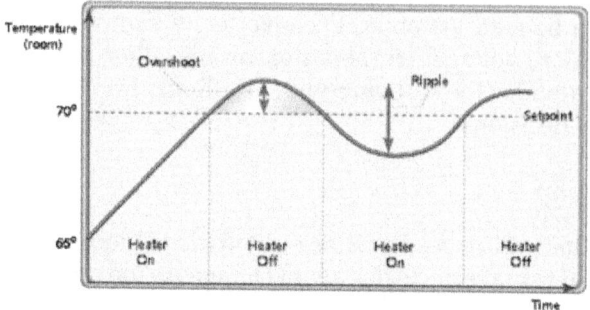

Fig. : Control example.

Proportional control is the primary alternative to on-off control. If the difference between the current plant output and its desired value (the current error) is large, the software should probably change the drive signal a lot. If the error is small, it should change it only a little. In other words, we always want a change like:

P * (desired -- current)

where P is a constant proportional gain set by the system's designer.

For example, if the drive signal uses PWM it can take any value between 0% and 100% duty cycle. If the signal on the drive is 20% duty cycle and the error remaining at the output is small, we may just need to tweak it to 18% or 19% to achieve the desired output at the plant.

If the proportional gain is well chosen, the time the plant takes to reach a new setpoint will be as short as possible, with overshoot (or undershoot) and oscillation minimized.

Unfortunately, proportional control alone is not sufficient in all control applications. One or more of the requirements for response time, overshoot, and oscillation may be impossible to fulfill at any proportional gain setting.

Differentiation

The biggest problem with proportional control alone is that you want to reach new desired outputs quickly and avoid overshoot and minimize ripple once you get there. Responding quickly suggests a high proportional gain; minimizing overshoot and oscillation suggests a small proportional gain. Achieving both at the same time may not be possible in all systems.

Fortunately, we do generally have (or can derive) information about the rate of change of the plant's output. If the output is changing rapidly, overshoot or undershoot may lie ahead. In that case, we can reduce the size of the change suggested by the proportional controller.

The rate of change of a signal is also known as its derivative. The derivative at the current time is simply the change in value from the previous sample to the current one. This implies that we should subtract a change of:

$$D * (current -- previous)$$

where D is a constant derivative gain. The only other thing we need to do is to save the previous sample in memory.

In practice, proportional-derivative (PD) controllers work well. The net effect is a slower response time with far less overshoot and ripple than a proportional controller alone.

Integration

A remaining problem is that PD control alone will not always settle exactly to the desired output. In fact, depending on the proportional gain, it's altogether possible that a PD controller will ultimately settle to an output value that is far from that desired.

The problem occurs if each individual error remains below the threshold for action by the proportional term. (Say the error is 3, $P = 1/8$, and integer math is used.) The derivative term won't help anything unless the output is changing. Something else needs to drive the plant toward the setpoint. That something is an integral term.

An integral is a sum over time, in this case the sum of all past errors in the plant output:

$$I * \sum_{t} \left(desired - current \right)$$

Even though the integral gain factor, I, is typically small, a persistent error will eventually cause the sum to grow large and the integral term to force a change in the drive signal.

In summary, on-off and proportional control are the two basic techniques of closed-loop control. However, derivative and/or integral terms are sometimes added to porportional controllers to improve qualitative properties of a particular plant's response. When all three terms are used together, the acronym used to describe the controller is PID.

OPEN- VS. CLOSED-LOOP CONTROL

Arguably the most ingenious tool of the control engineering profession is the feedback loop shown in the Basic Feedback Loop graphic. It consists of five fundamental elements:

- The process that is to be controlled
- An instrument with a sensor that measures the condition of the process
- A transmitter that converts the measurement into an electronic signal
- A controller that reads the transmitter's signal and decides whether or not the current condition of the process is acceptable, and
- An actuator functioning as the final control element that applies a corrective effort to the process per the controller's instructions.

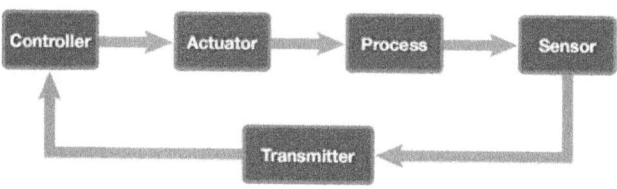

In a closed-loop control system, information flows around a feedback loop from the process to the sensor to the transmitter to the controller to the actuator and back to the process. This measure-decide-actuate sequence-known as closed-loop control-repeats as often as necessary until the desired process condition is achieved. Familiar examples include using a thermostat controlling a furnace to maintain the temperature in a room or cruise control to maintain the speed of a car.

But not all automatic control operations require feedback. A much larger class of control commands can be executed in an open-loop configuration without confirmation or further adjustment. Open-loop control is sufficient for predictable operations such as opening a door, starting a motor, or turning off a pump.

Continuous Closed-loop Control

Continuing the analysis, it is clear that all closed-loop operations are not alike. For a continuous process, a feedback loop attempts to maintain a process variable (or controlled variable) at a desired value known as the setpoint. The controller subtracts the latest measurement of the process variable from the setpoint to generate an error signal. The magnitude and duration of the error signal then determines the value of the controller's output or manipulated variable which in turn dictates the corrective efforts applied by the actuator.

For example, a car equipped with a cruise control uses a speedometer to measure and maintain the car's speed. If the car is traveling too slowly, the controller instructs the accelerator to feed more fuel to the engine. If the car is traveling too quickly, the controller lets up on the accelerator. The car is the process, the speedometer is the sensor, and the accelerator is the actuator.

The car's speed is the process variable. Other common process variables include temperatures, pressures, flow rates, and tank levels. These are all quantities that can vary constantly and can be measured at any time. Common actuators for manipulating such conditions include heating elements, valves, and dampers.

Discrete Closed-loop Control

For a discrete process, the variable of interest is measured only when a triggering event occurs, and the measure-decide-actuate sequence is typically executed just once for each event. For example, the human controller driving the car uses her eyes to measure ambient light levels at the beginning of each trip. If she decides that it's too dark to see well, she turns on the car's lights. No further adjustment is required until the next triggering event such as the end of the trip.

Feedback loops for discrete processes are generally much simpler than continuous control loops since discrete processes do not involve as much inertia. The driver controlling the car gets instantaneous results after turning on the lights, whereas the cruise control sees much more gradual results as the car slowly speeds up or slows down.

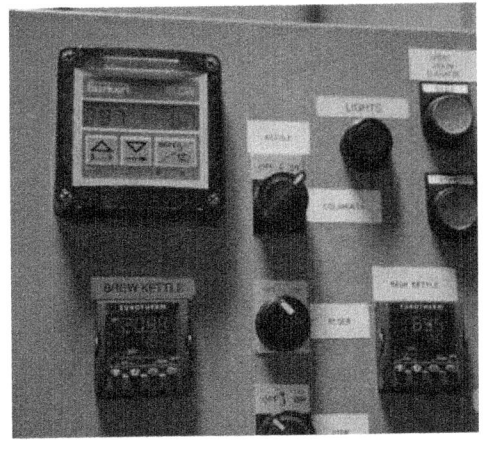

Inertia tends to complicate the design of a continuous control loop since a continuous controller typically needs to make a series of decisions before the results of its earlier efforts are completely evident. It has to anticipate the cumulative effects of its recent corrective efforts and plan future efforts accordingly. Waiting to see how each one turns out before trying another simply takes too long.

Open-loop Control

Open-loop controllers do not use feedback per se. They apply a single control effort when so commanded and assume that the desired results will be achieved. An open-loop controller may still measure the results of its commands: Did the door actually open? Did the motor actually start? Is the pump actually off? Generally, these actions are for safety considerations rather than as part of the control sequence.

Even closed-loop feedback controllers must operate in an open-loop mode on occasion. A sensor may fail to generate the feedback signal or an operator may take over the feedback operation in order to manipulate the controller's output manually.

Operator intervention is generally required when a feedback controller proves unable to maintain stable closed-loop control. For example, a particularly aggressive pressure controller may overcompensate for a drop in line pressure. If the controller then overcompensates for its overcompensation, the pressure may end up lower than before, then higher, then even lower, then even higher, *etc.* The simplest way to terminate such unstable oscillations is to break the loop and regain control manually.

There are also many applications where experienced operators can make manual corrections faster than a feedback controller can. Using their knowledge of the process' past behavior, operators can manipulate the process inputs now to achieve the desired output values later. A feedback controller, on the other hand, must wait until the effects of its latest efforts are measurable before it can decide on the next appropriate control action. Predictable processes with longtime constants or excessive dead time are particularly suited for open-loop manual control.

Open- and Closed-loop Control Combined

The principal drawback of open-loop control is a loss of accuracy. Without feedback, there is no guarantee that the control efforts applied to the process will actually have the desired effect. If speed and accuracy are both required, open-loop and closed-loop control can be applied simultaneously using a feedforward strategy. A feedforward controller uses a mathematical model of the process to make its initial control moves like an experienced operator would. It then measures the results of its open-loop efforts and makes additional corrections as necessary like a traditional feedback controller.

Feedforward is particularly useful when sensors are available to measure an impending disturbance before it hits the process. If its future effects on the process can be accurately predicted with the process model, the controller can take preemptive actions to counteract the disturbance as it occurs.

For example, if a car equipped with cruise control and radar could see a hill coming, it could begin to accelerate even before it begins to slow down. The car may not end up at the desired speed as it climbs the hill, but even that error can eventually be eliminated by the cruise controller's normal feedback control algorithm. Without the advance notice provided by the radar, the cruise controller wouldn't know that acceleration is required until the car had already slowed below the desired speed halfway up the hill.

Open- and Closed-loop Control in Parallel

Many automatic control systems use both open- and closed-loop control in parallel. Consider, for example, a brewery that ferments and bottles beer.

Brew kettles in a modern brewery rely on continuous closed-loop control to maintain prescribed temperatures and pressures while turning water and grain into fermentable mash.

A brewery's bottling line uses both discrete closed-loop control and open-loop control to fill and cap the individual bottles.

The conditions inside the brew kettles are maintained by closed-loop controllers using feedback loops that measure the temperature and pressure, then adjust steam flow into the kettle and flow pumps to compensate for out-of-spec conditions. Open-loop control is also required for one-time operations such as starting and stopping the mixer motors or opening and closing the steam lines to the heat exchangers.

Simultaneously, finished batches of beer are bottled using open-loop and discrete closed-loop control. A proximity sensor determines that a bottle is present before filling can begin, then a valve opens to fill each bottle until a level sensor determines that the bottle is full. In general, continuous closed-loop control applications require at least a few ancillary open-loop control operations, whereas many open-loop control applications require no feedback loops at all.

CONTROL STRATEGIES

A control strategy is a set of discrete and specific measures identified and implemented to achieve reductions in air pollution. These measures may vary by source type, such as stationary or mobile, as well as by the pollutant that is being targeted. The purpose of these measures is to achieve the air quality standard or goal. Costs and benefits are assessed in the development of the control strategy.

Control strategy development is the process of assessing specific abatement measures, management practices, or control technologies to determine the best combination of approaches to provide the emission reductions necessary to achieve the air quality standard or goal. Three primary considerations in designing an effective control strategy are:

(1) **Environmental:** factors such as equipment locations, ambient air quality conditions, adequate utilities (*i.e.*, water for scrubbers), legal requirements, noise levels, and the contribution of the control system as a pollutant;

(2) **Engineering:** factors such as contaminant characteristics (abrasiveness, toxicity, *etc.*), gas stream characteristics, and performance characteristics of the control system; and

(3) **Economic:** factors such as capital cost, operating costs, equipment maintenance, and the lifetime of the equipment. Air pollution officials should also consider pollution prevention which includes eliminating as much of the pollution emissions as possible at the source, substituting raw (and less toxic) materials, considering alternative manufacturing processes, and improving process control measures.

Controls on major stationary, mobile, and area sources are part of a successful control strategy. These controls should utilize reasonably available control technology. Examples include controls on volatile organic compounds from solvent and paint usage as well as controls on nitrogen oxide emissions from combustion units. For mobile sources, examples include tighter emission controls for vehicles and low-sulfur fuel standards. For major stationary sources, it is beneficial to issue permits including emission limitations for any major sources, new and existing. The basic types of emission control technology are mechanical collectors, wet scrubbers, bag houses, electrostatic precipitators, combustion systems (thermal oxidizers), condensers, absorbers, adsorbers, and biological degradation. The selection procedure should be based on the environmental, engineering, and economic considerations described above.

National or local governments new to the air quality management process should focus on obvious sources of air pollution and the quickest means of control - more sophisticated and comprehensive strategies can be developed over time. Innovative strategies such as emissions trading, banking, and emissions caps can be incorporated as a further refinement as the strategy continues. These strategies may be used in addition to the "command-and-control" type regulations which have traditionally been used by air pollution control agencies. Local and regional control measures and are both necessary for a successful strategy.

A control strategy developed by a local government may include locally appropriate measures, as well as control measures that the national government mandates be implemented nation-wide. Successful control strategies are usually adopted into a regulatory program with implementation deadlines and mechanisms for enforcement. Different control measures may be mandated at different levels of government, from local to provincial, state level to national. In general, regulations established at the national level tend to have the most benefit while minimizing boundary and competition issues. The goal for all control strategies is to achieve real and measurable emission reductions.

How do I Develop a Control Strategy?

There are four main steps in developing a control strategy.

(1) **Determine priority pollutants**. The pollutants of concern for your location should be based on health effects and the severity of the air quality problem.

(2) **Identify control measures.** For specific source categories, choose the appropriate controls based on the priority pollutants identified. A good source for control technologies is U.S. EPA's Clean Air Technology Center. Also, the National Association of Clean Air Agencies (NACAA) has developed several documents that provide a menu of control options. To order any of these documents (listed below), visit the NACAA website.

- Controlling Particulate Matter Under the Clean Air Act: A Menu of Options

- Controlling Nitrogen Oxides Under the Clean Air Act: A Menu of Options

- Meeting the VOC 15% Rate-of-Progress Requirement Under the Clean Air Act: A Menu of Options

- Toxic Air Pollutants: State and Local Regulatory Strategies

(3) **Incorporate the control measures into a plan.** Using the control measures identified, create a written plan with implementation dates to formalize the strategy. It is important to adopt a regulatory program and include it in the plan so that control measures will be enforceable.

(4) **Involve the public.** As with the other management activities related to the AQM process, it is critical to contact the regulated community and other affected parties, as the public should be consulted as part of the strategy development process. This early consultation reduces later challenges and streamlines implementation.

In the U.S., individual states are responsible for this process and they develop their own plan (called the State Implementation Plan, or SIP) based on the air quality issues that are of concern in their region. The SIP is the federally-enforceable plan that identifies how that state will attain and/or maintain the primary and secondary National Ambient Air Quality Standards (NAAQS) set forth in the Clean Air Act (CAA). Each state is required to have a SIP which contains the control measures and strategies developed through a public process and is formally adopted by the state and submitted to the U.S. EPA. These plans cover the control of all air pollution issues from industry and automobile emissions to open burning and more. Another example is given in the SIP Plan Summaries for the New England region of the U.S.

STRATEGIC CONTROL

Strategic control is a term used to describe the process used by organizations to control the formation and execution of strategic plans; it is a specialised form of management control, and differs from other forms of management control (in particular from operational control) in respects of its need to handle *uncertainty* and *ambiguity* at various points in the control process.

Strategic control is also focused on **the achievement of future goals**, rather than the evaluation of past performance. Vis:

The purpose of control at the strategic level is not to answer the question:' 'Have we made the right strategic choices at some time in the past?" but rather "How well are we doing now and how well will we be doing in the immediate future for which reliable information is available?" The point is not to bring to light past errors but to identify needed corrections to steer the corporation in the desired direction. And this determination must be made with respect to currently desirable long-range goals and not against the goals or plans that were established at some time in the past.

As with other control processes, strategic control processes are at their core cybernetic in nature: using one or more 'closed loop' controls to ensure that any observed deviations from expected activity or outcomes are highlighted to managers who can then intervene to correct / adjust the organisation's future activities. John Preble noted the need for these controls to be 'forward looking' when used to control strategy, to give controls that are "future-directed and anticipatory".

Strategic control systems cannot "...wait for a strategy to be executed before getting any feedback on how well it is working. Since this might take several years..."

A related concern for strategic control processes is the amount of time and effort required for the process to work: if either is too great the process will either be ineffective or be ignored by the organisation.

Various authors have proposed that all strategic control systems necessarily comprise a small set of standard elements, the absence of any one of which makes strategic control impossible to achieve (*e.g.* Goold & Quinn, Muralidharan). The four elements proposed by Muralidharan are:

- the articulation of the *strategic outcomes* being sought
- the description of the *strategic activities* to be carried out (attached to specific managed resources) in pursuit of the required outcomes
- the definition of a *method to track progress* made by the against these two elements (usually via the monitoring of a small number of performance measures and associated target values)
- the identification of an *effective intervention mechanism* that would allow observers (usually the organisation's managers) to change / correct / adjust the organisation's activities when targets are not achieved

These elements imply an active involvement by senior managers in the determination of the strategic activities pursued by the component parts of an organisation, and this has lead some to observe that strategic control is most effective in organisations that focus on a single market or area of activity. In organisations undertaking a mix of diverse / unrelated activities (*e.g.* traditional conglomerates) simpler forms of financial control are more common and perhaps more effective.

History

Although *control* was one of the six 'functions of management'. listed by Henri Fayol in 1917, the idea of strategic control as a distinct activity does not appear in the management literature until the late 1970s (*e.g.* "Strategic Control: a new task for top management" by J H Horovitz, which was published in 1979, is a candidate for first paper to explicitly discuss the topic), but the first definition of strategic control in a form consistent with modern usage of the term is probably in a paper by Reufli and Sarrazin published in 1981

As Reufli and Sarrazin observed, the key issue with strategic control mechanisms is the need to dealing with uncertainty and ambiguity. A landmark study by Michael Goold and Andrew Campbell identified that a variety of control methods are used across a continuum ranging from purely *financial controls* at one extreme, through to detailed *strategic planning* systems at the other. They observed a series of trade-offs between these extremes - *financial controls* being simpler and therefore cheaper and more flexible to operate, but providing less scope for coordination between components of an organisation, *strategic planning* being time-consuming and expensive to operate, but providing the greatest scope to push for maximum strategic advantage. In the middle of this range, Goold and Campbell described *strategic control* as allowing firms to "balance competitive and financial ambitions". This idea of a spectrum of control has since been widely adopted.

Links to Other Tools

Although *strategic control* is a general management topic rather than a prescriptive tool, its reliance on feedback on organisational performance has resulted in a long association with performance management tools such as the balanced scorecard and its derivatives such as the performance measurement, and with strategy implementation frameworks such as that proposed by Hrebiniak and Joyce.

CONTROL LOGIC

Control logic is a key part of a software [computer program | program] that controls the operations of the program. The control logic responds to commands from the user, and it also acts on its own to perform automated tasks that have been structured into the program.

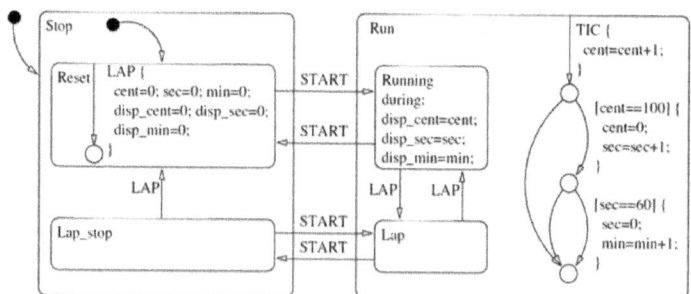

Fig. : Model of a simple stopwatch.

Control logic can be modeled using a state diagram, which is a form of hierarchical state machine. These state diagrams can also be combined with flow charts to provide a set of computational semantics for describing complex control logic. This mix of state diagrams and flow charts is illustrated in the figure, which shows the control logic for a simple stopwatch. The control logic takes in commands from the user, as represented by the event named "START", but also has automatic recurring sample time events, as represented by the event named "TIC".

NEGATIVE FEEDBACK

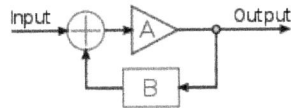

Fig. : A simple negative feedback system descriptive, for example, of some electronic amplifiers. The feedback is negative if the loop gain AB is negative.

Negative feedback occurs when some function of the output of a system, process, or mechanism is fed back in a manner that tends to reduce the fluctuations in the output, whether caused by changes in the input or by other disturbances.

Whereas positive feedback tends to lead to instability via exponential growth, oscillation or chaotic behavior, negative feedback generally promotes stability. Negative feedback tends to promote a settling to equilibrium, and reduces the effects of perturbations. Negative feedback loops in which just the right amount of correction is applied with optimum timing can be very stable, accurate, and responsive.

Negative feedback is widely used in mechanical and electronic engineering, but it also occurs naturally within living organisms, and can be seen in many other fields from chemistry and economics to physical systems such as the climate. General negative feedback systems are studied in control systems engineering.

Examples

Mercury thermostats (circa 1600) using expansion and contraction of columns of mercury in response to temperature changes were used in negative feedback systems to control vents in furnaces, maintaining a steady internal temperature.

In the invisible hand of the market metaphor of economic theory (1776), reactions to price movements provide a feedback mechanism to match supply and demand.

In centrifugal governors (1788), negative feedback is used to maintain a near-constant speed of an engine, irrespective of the load or fuel-supply conditions.

In a Steering engine (1866), power assistance is applied to the rudder with a feedback loop, to maintain the direction set by the steersman.

In servomechanisms, the speed or position of an output, as determined by a sensor, is compared to a set value, and any error is reduced by negative feedback to the input.

In audio amplifiers, negative feedback reduces distortion, minimises the effect of manufacturing variations in component parameters, and compensates for changes in characteristics due to temperature change.

In analog computing feedback around operational amplifiers is used to generate mathematical functions such as addition, subtraction, integration, differentiation, logarithm, and antilog functions.

In a phase locked loop (1932) feedback is used to maintain a generated alternating waveform in a constant phase to a reference signal. In many implementations the generated waveform is the output, but when used as a demodulator in a FM radio receiver, the error feedback voltage serves as the demodulated output signal. If there is a frequency divider between the generated waveform and the phase comparator, the device acts as a frequency multiplier.

In organisms, feedback enables various measures (*e.g.* body temperature, or blood sugar level) to be maintained within a desired range by homeostatic processes.

History

Negative feedback as a control technique may be seen in the refinements of the water clock introduced by Ktesibios of Alexandria in the 3rd century BCE. Self-regulating mechanisms have existed since antiquity, and were used to maintain a constant level in the reservoirs of water clocks as early as 200 BCE.

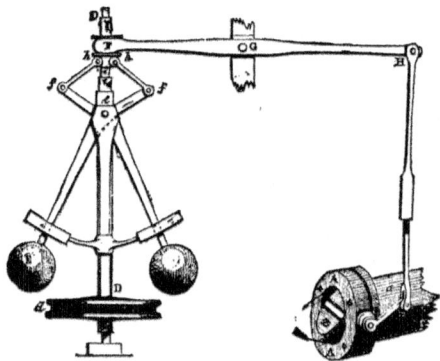

Fig. : The fly-ball governor is an early example of negative feedback.

Negative feedback was implemented in the 17th Century. Cornelius Drebbel had built thermostatically-controlled incubators and ovens in the early 1600s, and centrifugal governors were used to regulate the distance and pressure between millstones in windmills. James Watt patented a form of governor in 1788 to control the speed of his steam engine, and James Clerk Maxwell in 1868 described "component motions" associated with these governors that lead to a decrease in a disturbance or the amplitude of an oscillation.

The term "feedback" was well established by the 1920s, in reference to a means of boosting the gain of an electronic amplifier. Friis and Jensen described this action as "positive feedback" and made passing mention of a contrasting "negative feed-back action" in 1924. Harold Stephen Black came up with the idea of using negative feedback in electronic amplifiers in 1927, submitted a patent application in 1928, and detailed its use in his paper of 1934, where he defined negative feedback as a type of coupling that *reduced* the gain of the amplifier, in the process greatly increasing its stability and bandwidth.

Karl Küpfmüller published papers on a negative-feedback-based automatic gain control system and a feedback system stability criterion in 1928.

Nyquist and Bode built on Black's work to develop a theory of amplifier stability.

Early researchers in the area of cybernetics subsequently generalized the idea of negative feedback to cover any goal-seeking or purposeful behavior.

All purposeful behavior may be considered to require negative feed-back. If a goal is to be attained, some signals from the goal are necessary at some time to direct the behavior.

Cybernetics pioneer Norbert Wiener helped to formalize the concepts of feedback control, defining feedback in general as "the chain of the transmission and return of information", and negative feedback as the case when:

The information fed back to the control center tends to oppose the departure of the controlled from the controlling quantity...

While the view of feedback as any "circularity of action" helped to keep the theory simple and consistent, Ashby pointed out that, while it may clash with definitions that require a "materially evident" connection, "the exact definition of feedback is nowhere important". Ashby pointed out the limitations of the concept of "feedback":

The concept of 'feedback', so simple and natural in certain elementary cases, becomes artificial and of little use when the interconnections between the parts become more complex...Such complex systems cannot be treated as an interlaced set of more or less independent feedback circuits, but only as a whole. For understanding the general principles of dynamic systems, therefore, the concept of feedback is inadequate in itself. What is important is that complex systems, richly cross-connected internally, have complex behaviors, and that these behaviors can be goal-seeking in complex patterns.

To reduce confusion, later authors have suggested alternative terms such as *degenerative, self-correcting, balancing,* or *discrepancy-reducing* in place of "negative".

Overview

In many physical and biological systems, qualitatively different influences can oppose each other. For example, in biochemistry, one set of chemicals drives the system in a given direction, whereas another set of chemicals drives it in an opposing direction. If one or both of these opposing influences are non-linear, equilibrium point(s) result.

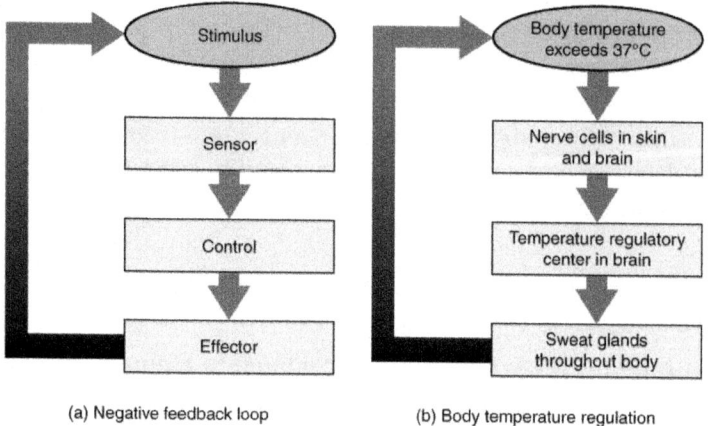

(a) Negative feedback loop   (b) Body temperature regulation

Fig. : Feedback loops in the human body.

In biology, this process (in general, biochemical) is often referred to as homeostasis; whereas in mechanics, the more common term isequilibrium.

In engineering, mathematics and the physical, and biological sciences, common terms for the points around which the system gravitates include: attractors, stable states, eigenstates/eigenfunctions, equilibrium points, and setpoints.

In control theory, *negative* refers to the sign of the multiplier in mathematical models for feedback. In delta notation, $-\Delta$ output is added to or mixed into the input. In multivariate systems, vectors help to illustrate how several influences can both partially complement and partially oppose each other.

Some authors, in particular with respect to modelling business systems, use *negative* to refer to the reduction in difference between the desired and actual behavior of a system. In a psychology context, on the other hand, *negative* refers to the valence of the feedback – attractive versus aversive, or praise versus criticism.

In contrast, positive feedback is feedback in which the system responds so as to increase the magnitude of any particular perturbation, resulting in amplification of the original signal instead of stabilization. Any system in which there is positive feedback together with a gain greater than one will result in a runaway situation. Both positive and negative feedback require a feedback loop to operate.

However, negative feedback systems can still be subject to oscillations. This is caused by the slight delays around any loop. Due to these delays the feedback signal of some frequencies can arrive one half cycle later which will have a similar effect to positive feedback and these frequencies can reinforce themselves and grow over time. This problem is often dealt with by attenuating or changing the phase of the problematic frequencies. Unless the system naturally has sufficient damping, many negative feedback systems have low pass filters or dampers fitted.

Some Specific Implementations

There are a large number of different examples of negative feedback and some are discussed below.

Error-controlled Regulation

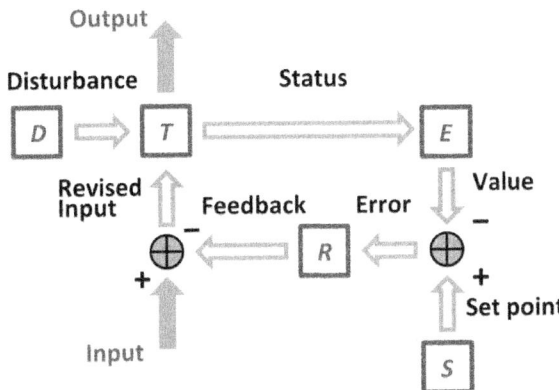

Fig. : A regulator R adjusts the input to a system T so the monitored essential variables E are held to set-point values S that result in the desired system output despite disturbances D.

One use of feedback is to make a system (say *T*) self-regulating to minimize the effect of a disturbance (say *D*). Using a negative feedback loop, a measurement of some variable (for example, a process variable, say *E*) is subtracted from a required value (the 'set point') to estimate an operational error in system status, which is then used by a regulator (say *R*) to reduce the gap between the measurement and the required value. The regulator modifies the input to the system *T* according to its interpretation of the error in the status of the system. This error may be introduced by a variety of possible disturbances or 'upsets', some slow and some rapid. The regulation in such systems can range from a simple 'on-off' control to a more complex processing of the error signal.

It may be noted that the physical form of the signals in the system may change from point to point. So, for example, a change in weather may cause a disturbance to the *heat* input to a house (as an example of the system *T*) that is monitored by a thermometer as a change in *temperature* (as an example of an 'essential variable' *E*), converted by the thermostat (a 'comparator') into an *electrical* error in status compared to the 'set point' *S*, and subsequently used by the regulator (containing a 'controller' that commands *gas* control valves and an ignitor) ultimately to change the *heat* provided by a furnace (an 'effector') to counter the initial weather-related disturbance in heat input to the house.

Error controlled regulation is typically carried out using a Proportional-Integral-Derivative Controller (PID controller). The regulator signal is derived from a weighted sum of the error signal, integral of the error signal, and derivative of the error signal. The weights of the respective components depend on the application.

Mathematically, the regulator signal is given by:

$$MV(t) = K_p \left(e(t) + \frac{1}{T_i} \int_0^t e(\tau)\, d\tau + T_d \frac{d}{dt} e(t) \right)$$

where

T_i is the *integral time*

T_d is the *derivative time*

Negative Feedback Amplifier

The negative feedback amplifier was invented by Harold Stephen Black at Bell Laboratories in 1927, and granted a patent in 1937.

There are many advantages to feedback in amplifiers. In design, the type of feedback and amount of feedback are carefully selected to weigh and optimize these various benefits.

Though negative feedback has many advantages, amplifiers with feedback can oscillate. See the article on step response. They may even exhibit instability. Harry Nyquist of Bell Laboratories proposed the Nyquist stability criterion and the Nyquist plot that identify stable feedback systems, including amplifiers and control systems.

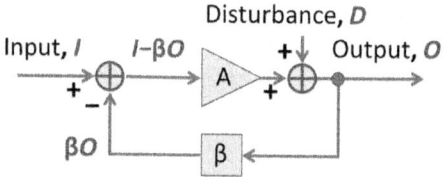

Fig. : Negative feedback amplifier with external disturbance.
The feedback is negative if $\beta A > 0$.

The figure shows a simplified block diagram of a negative feedback amplifier.

The feedback sets the overall (closed-loop) amplifier gain at a value:

$$\frac{O}{I} = \frac{A}{1 + \beta A} \approx \frac{1}{\beta},$$

where the approximate value assumes $\beta A \gg 1$. This expression shows that a gain greater than one requires $\beta < 1$. Because the approximate gain $1/\beta$ is independent of the open-loop gain A, the feedback is said to 'desensitize' the closed-loop gain to variations in A (for example, due to manufacturing variations between units, or temperature effects upon components), provided only that the gain A is sufficiently large. In this context, the factor $(1 + \beta A)$ is often called the 'desensitivity factor', and in the broader context of feedback effects that include other matters like electrical impedance and bandwidth, the 'improvement factor'.

If the disturbance D is included, the amplifier output becomes:

$$O = \frac{AI}{1+\beta A} + \frac{D}{1+\beta A},$$

which shows that the feedback reduces the effect of the disturbance by the 'improvement factor' $(1 + \beta A)$. The disturbance D might arise from fluctuations in the amplifier output due to noise and nonlinearity (distortion) within this amplifier, or from other noise sources such as power supplies.

The difference signal I–βO at the amplifier input is sometimes called the "error signal". According to the diagram, the error signal is:

$$\text{Error signal} = I - \beta O = I\left(1 - \beta \frac{O}{I}\right) = \frac{I}{1+\beta A} - \frac{\beta D}{1+\beta A}.$$

From this expression, it can be seen that a large 'improvement factor' (or a large loop gain βA) tends to keep this error signal small.

Although the diagram illustrates the principles of the negative feedback amplifier, modeling a real amplifier as a unilateral forward amplification block and a unilateral feedback block has significant limitations.

Operational Amplifier Circuits

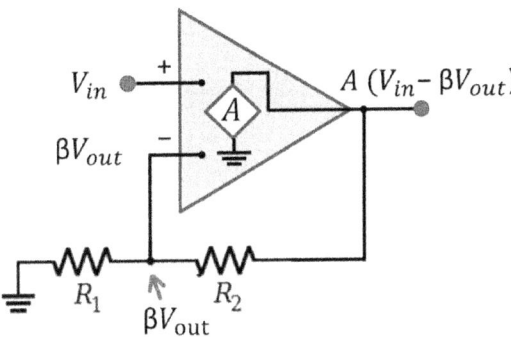

Fig. : A feedback voltage amplifier using an op amp with finite gain but infinite input impedances and zero output impedance.

The operational amplifier was originally developed as a building block for the construction of analog computers, but is now used almost universally in all kinds of applications including audio equipment and control systems.

Operational amplifier circuits typically employ negative feedback to get a predictable transfer function. Since the open-loop gain of an op-amp is extremely large, a small differential input signal would drive the output of the amplifier to one rail or the other in the absence of negative feedback. A simple example of the use of feedback is the op-amp voltage amplifier shown in the figure.

The idealized model of an operational amplifier assumes that the gain is infinite, the input impedance is infinite, output resistance is zero, and input offset currents and voltages are zero. Such an ideal amplifier draws no current from the resistor divider. Ignoring dynamics (transient effects and propagation delay), the infinite gain of the ideal op-amp means this feedback circuit drives the voltage difference between the two op-amp inputs to zero. Consequently, the voltage gain of the circuit in the diagram, assuming an ideal op amp, is the reciprocal of feedback voltage division ratio β:

$$V_{out} = \frac{R_1 + R_2}{R_1} V_{in} = \frac{1}{\beta} V_{in}.$$

A real op-amp has a high but finite gain A at low frequencies, decreasing gradually at higher frequencies. In addition, it exhibits a finite input impedance and a non-zero output impedance. Although practical op-amps are not ideal, the model of an ideal op-amp often suffices to understand circuit operation at low enough frequencies.

Mechanical Engineering

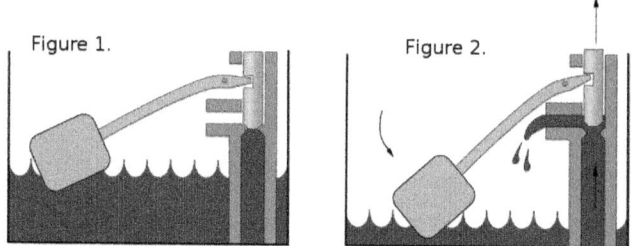

Fig. : The ballcock or float valve uses negative feedback to control the water level in a cistern.

An example of the use of negative feedback control is the ballcock control of water level. In modern engineering, negative feedback loops are found in fuel injection systems and carburettors. Similar control mechanisms are used in heating and cooling systems, such as those involving air conditioners, refrigerators, or freezers.

Biology and Chemistry

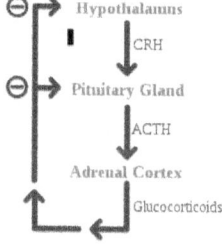

Fig. : Control of endocrine hormones by negative feedback.

Some biological systems exhibit negative feedback such as the barore-flex in blood pressure regulation and erythropoiesis. Many biological process (*e.g.,* in the human anatomy) use negative feedback. Examples of this are numerous, from the regulating of body temperature, to the regulating of blood glucose levels. The disruption of feedback loops can lead to undesirable results: in the case of blood glucose levels, if negative feedback fails, the glucose levels in the blood may begin to rise dramatically, thus resulting in diabetes.

For hormone secretion regulated by the negative feedback loop: when gland X releases hormone X, this stimulates target cells to release hormone Y. When there is an excess of hormone Y, gland X "senses" this and inhibits its release of hormone X. As shown in the figure, most endocrine hormones are controlled by a physi-ologic negative feedback inhibition loop, such as the glucocorticoids secreted by theadrenal cortex. The hypothalamus secretes corticotropin-releasing hormone (CRH), which directs the anterior pituitary gland to secrete adrenocorticotropic hormone (ACTH). In turn, ACTH directs the adrenal cortex to secrete glucocorti-coids, such as cortisol. Glucocorticoids not only perform their respective functions throughout the body but also negatively affect the release of further stimulating secretions of both the hypothalamus and the pituitary gland, effectively reducing the output of glucocorticoids once a sufficient amount has been released.

Self-organization

Self-organization is the capability of certain systems "of organizing their own behavior or structure". There are many possible factors contributing to this capacity, and most often positive feedback is identified as a possible contributor. However, negative feedback also can play a role.

FEEDBACK CONTROL OF HORMONE PRODUCTION

Feedback circuits are at the root of most control mechanisms in physiology, and are particularly prominent in the endocrine system. Instances of positive feedback certainly occur, but negative feedback is much more common.

Negative feedback is seen when the output of a pathway inhibits inputs to the pathway. The heating system in your home is a simple negative feedback circuit. When the furnace produces enough heat to elevate temperature above the set point of the thermostat, the thermostat is triggered and shuts off the furnace (heat is feeding back negatively on the source of heat). When temperature drops back below the set point, negative feedback is gone, and the furnace comes back on.

Feedback loops are used extensively to regulate secretion of hormones in the hypothalamic-pituitary axis. An important example of a negative feedback loop is seen in control of thyroid hormone secretion. The thyroid hormones thyroxine and triiodothyronine ("T4 and T3") are synthesized and secreted by thyroid glands and affect metabolism throughout the body. The basic mechanisms for control in this system are:

- Neurons in the hypothalamus secrete thyroid releasing hormone (TRH), which stimulates cells in the anterior pituitary to secrete thyroid-stimulating hormone (TSH).

- TSH binds to receptors on epithelial cells in the thyroid gland, stimulating synthesis and secretion of thyroid hormones, which affect probably all cells in the body.

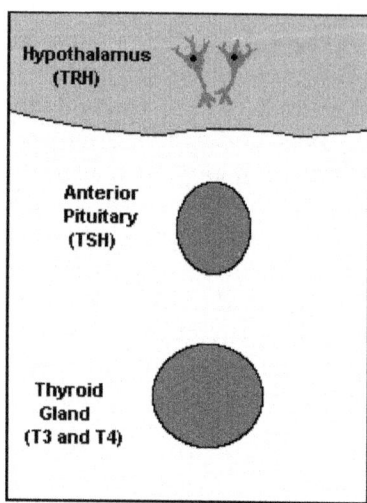

- When blood concentrations of thyroid hormones increase above a certain threshold, TRH-secreting neurons in the hypothalamus are inhibited and stop secreting TRH. *This is an example of "negative feedback".*

Inhibition of TRH secretion leads to shut-off of TSH secretion, which leads to shut-off of thyroid hormone secretion. As thyroid hormone levels decay below the threshold, negative feedback is relieved, TRH secretion starts again, leading to TSH secretion.

Another type of feedback is seen in endocrine systems that regulate concentrations of blood components such as glucose. Drink a glass of milk or eat a candy bar and the following (simplified) series of events will occur:

- Glucose from the ingested lactose or sucrose is absorbed in the intestine and the level of glucose in blood rises.

- Elevation of blood glucose concentration stimulates endocrine cells in the pancreas to release insulin.

- Insulin has the major effect of facilitating entry of glucose into many cells of the body - as a result, blood glucose levels fall.

- When the level of blood glucose falls sufficiently, the stimulus for insulin release disappears and insulin is no longer secreted.

Numerous other examples of specific endocrine feedback circuits are presented in the sections on specific hormones or endocrine organs.

Hormone Profiles: Concentrations Over Time

One important consequence of the feedback controls that govern hormone concentrations and the fact that hormones have a limited lifespan or halflife is that most hormones are secreted in "pulses". The following graph depicts concentrations of luteinizing hormone in the blood of a female dog over a period of 8 hours, with samples collected every 15 minutes:

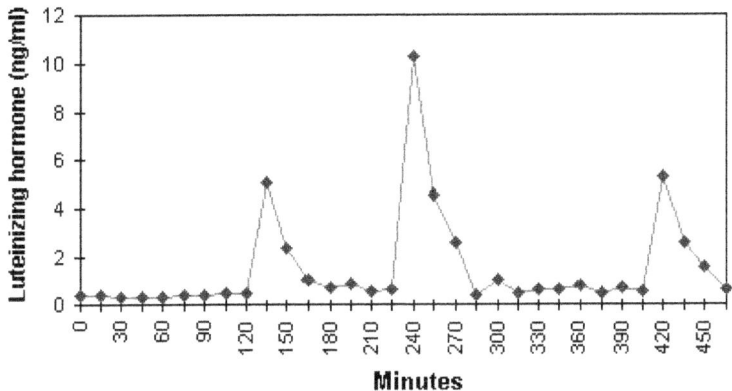

The pulsatile nature of luteinizing hormone secretion in this animal is evident. Luteinizing hormone is secreted from the anterior pituitary and critically involved in reproductive function; the frequency and amplitude of pulses are quite different at different stages of the reproductive cycle.

With reference to clinical endocrinology, examination of the graph should also demonstrate the caution necessary in interpreting endocrine data based on isolated samples.

A pulsatile pattern of secretion is seen for virtually all hormones, with variations in pulse characteristics that reflect specific physiologic states. In addition to the short-term pulses discussed here, longer-term temporal oscillations or endocrine rhythms are also commonly observed and undoubtedly important in both normal and pathologic states.

CONTROL SYSTEMS AND HOMEOSTASIS

Sometimes we act in response to a specific stimulus. When an object approaches the cornea of the eye, the lid closes to protect the cornea from damage. When the tendon of a muscle is tapped, the muscle contracts. In these cases it is easy to identify the stimulus and the response. Other times it is not easy to identify what caused a particular action. Sometimes the actions are part of some continuous activity. The body regulates its temperature continuously. It may increase or decrease its temperature when it finds that it is too cold or too hot. In this case, temperature is being regulated by a **control system**, and the control is called **homeostasis**.

Open Loop System

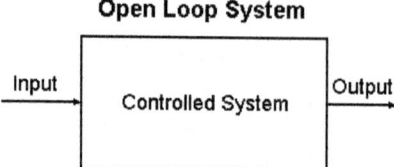

Fig. : An open loop system. The controlled system is indicated by the rectangle. Also indicated are the input to the system and its output.

Control Systems

Engineers refer to the first kind of action, a simple response to a stimulus, as an **open loop system**. It is useful to draw a diagram of how such a system works. Such a diagram is shown in Figure. The system being controlled is contained within the box marked "controlled system." That might be the system that controls the eyelid, for example. There is an input to the system--the approaching object--and an output from the system--the closing of the lid. The system simply responds to the input.

Closed Loop System

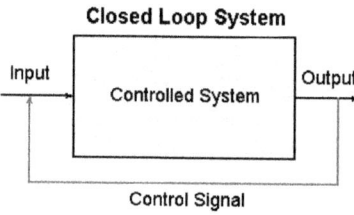

Fig. : A closed loop control system. The controlled system is indicated by the rectangle. Also indicated are the input, output and control signal.

On the other hand, the system that controls body temperature is called a **closed loop system**. The diagram for such a system is shown in Figure. There are also a controlled system, input and output in this kind of system, but, in addition, there is something that senses the output of the system and effects some change in the input. It is this connection that makes this a "closed loop" system.

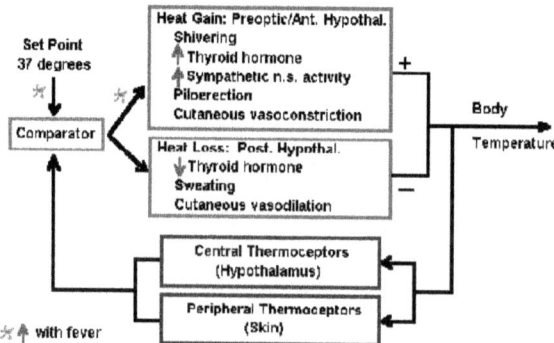

Fig. : Block diagram of the system for control of body temperature.

A diagram of the system that controls body temperature is shown in Figure. Somewhere in the brain, perhaps the hypothalamus, the optimum temperature of the body (**set point**) is stored. That information is continuously available to some structure, we call the comparator. The comparator sends signals to

- Heat gain mechanisms in the preoptic area or anterior hypothalamus leading to
  - o shivering
  - o increased thyroid hormone output
  - o increased activity in the sympathetic nervous system
  - o piloerection
  - o cutaneous vasoconstriction
- Heat loss mechanisms in the posterior hypothalamus leading to
  - o decreased thyroid hormone output
  - o sweating
  - o cutaneous vasodilation

The output of these mechanisms will end as either a net increase or a net decrease in body temperature. The body temperature is sensed by thermal receptors (thermoceptors) in the brain and peripherally in the body, and the value is sent to the comparator where it is compared with the set point. If the value is less than the set point, then signals go mainly to the heat gain mechanisms; if it is greater than the set point, then they go mainly to the heat loss mechanisms. In this way, body temperature is constantly sensed and maintained constant (*i.e.*, homeostasis).

Feedback Control

In a **feedback control system**, the output is sensed and this information is used at an earlier point in the system--it feeds back. Actually, Figure illustrates feedback control.

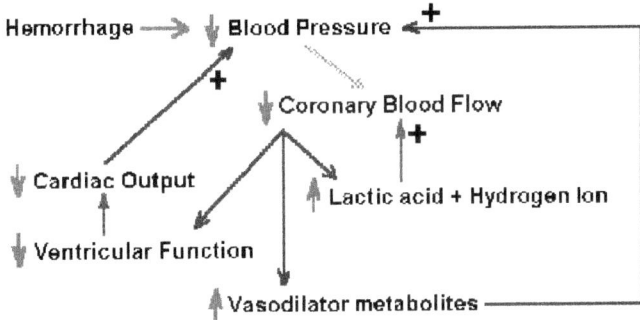

Fig. : Schematic diagram of the positive feedback system that is activated by hemorrhage. Like most positive feedback systems, this one is unstable, and if unchecked, will result in death.

Positive Feedback Systems

In a **positive feedback system**, the feedback is used to increase the size of the input. By nature, such systems are unstable, and they are most often associated with pathological conditions. An example of a positive feedback system is shown in Figure. In this diagram, hemorrhage leads to a decrease in blood pressure, which, it turn, leads to a decrease in flow in coronary arteries. The consequences of the decreased flow are

- increased lactic acid and hydrogen ion accumulation, which lead to further decrease in coronary blood flow

- increased vasodilator metabolites, which lead to further decreased blood pressure

- decreased contraction of the ventricles of the heart, which leads to decreased cardiac output and further decreased blood pressure

Clearly, none of these consequences is good. Several passages through this system will lead to excessive decrease in blood pressure and death. This is a positive feedback system because all of the consequences tend to increase the effect of the hemorrhage in lowering blood pressure.

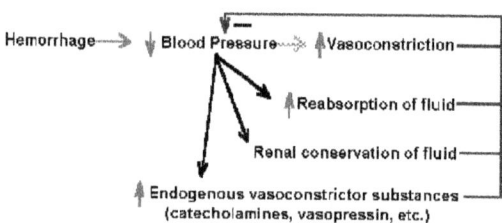

Fig. : Schematic diagram of the negative feedback system that is activated by hemorrhage.

Negative Feedback System

In a **negative feedback system**, the feedback is used to decrease the size of the input. These systems are usually stable, and they are associated with beneficial regulation of physiological parameters. An example of a negative feedback system is shown in Figure. In this diagram, hemorrhage leads to decreased blood pressure, which in turn leads to :

- increased reabsorption of fluid

- increased constriction of blood vessels

- increased renal conservation of fluid

- increased endogenous vasocontrictor substances, such as catecholamines and vasopressin

All of these lead to increased blood pressure. This consequence counters the effect of the initial hemorrhage and is, therefore, beneficial. This is a negative feedback system because all of the consequences tend to decrease the effect of the hemorrhage in lowering blood pressure.

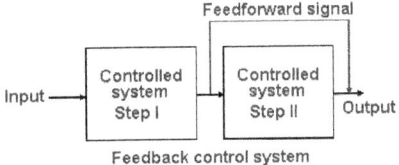

Fig. : Block diagram of a feedforward control system.

FEEDFORWARD SYSTEMS

In a feedforward system, the output of one stage of the processing of the control system is sent to a later stage of the process to affect later activity. The diagram for such a process is shown in Figure. An example of a feedforward system is the preadaptation for exercise, changing the activity of postural muscles and of the vascular system in order to ready the body for the movement when it occurs. Moving the arm laterally shifts the center-of-gravity laterally, and the person would be in danger of falling over were not compensations made in the postural musculature to prepare for the movement.

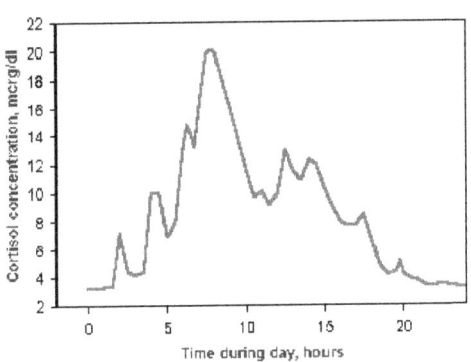

Fig. : Example of a diurnal rhythm. Cortisol concentrations in blood are controlled by a system that functions on a 24-hour cycle.

CHRONOTROPIC CONTROL

The time-frame over which a parameter is controlled can be quite variable, ranging from seconds to years. Here are some examples of different ones.

Diurnal Rhythms

Many parameters are regulated over a period approximating one day or 24 hours. They often are timed or "synched" by sleep-wakefulness or light-dark cycles. For example, the hormone cortisol, which is made by the adrenal gland and has important functions in metabolism of proteins, carbohydrates and fats, is controlled on a 24-hour cycle. Maximum concentration in blood are achieved between 7 and 8 am each day, with a nadir about midnight. This pattern can be seen in Figure.

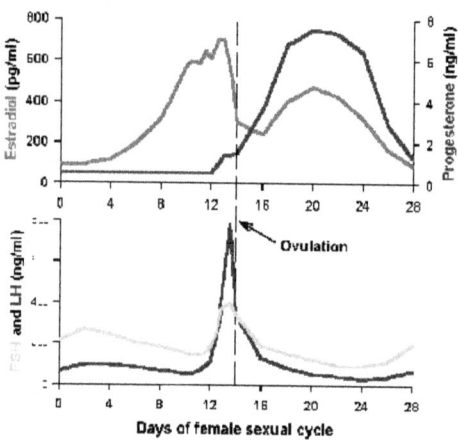

Fig. : Example of a lunar rhythm. Control of the ovulatory cycle in the human female occurs over an approximately 30-day period.

Lunar Rhythms

Some parameters are regulated over a period approximating one month or 30 days. Some appear to be tied to phases of the moon, but for others the synchronizing event is unclear. The ovulatory cycle in the human female is an example of such a lunar rhythm. This is a complicated mechanism but it suffices for our purposes now to say that ovulation occurs in the middle of the cycle and is triggered by increases in luteinizing hormone (LH). This can be seen in the center of the lower graph of Figure. The events depicted in the diagram are replayed every month during the reproductive span of the human female.

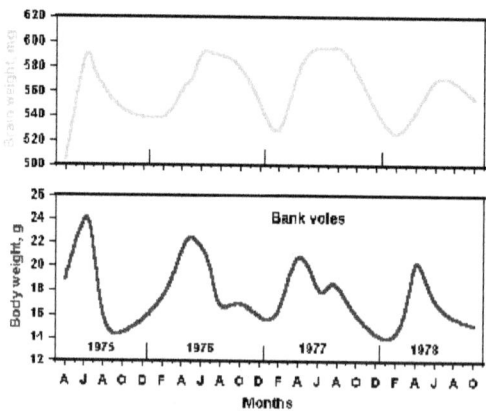

Fig. : Brain (upper) and body (lower) weights in the bank vole are controlled with a period of about 1 year.

Seasonal Rhythms

Many events are regulated with a period approximating one year. These could be synchronized by changes in temperature, sunlight or tides, and often the triggering stimulus has not been identified with certainty. An example of such a seasonal rhythm is shown in Figure. Here the brain weights (upper graph) and body weights (lower graph) have been measured on a monthly basis over a period of 4 years. Clearly, both parameters are at a maximum in March through May and at minimums in January through February. These measurements were made in bank voles, which have definite annual breeding cycles. Perhaps they would not occur in humans, which lack such cycles, but undoubtedly some other parameters would have seasonal rhythms in humans.

Developmental Rhythms

Some events are controlled on a life- time basis. For example, puberty occurs only once per individual. The same may be true for peak intellectual and physical performance. No one knows what triggers these events or why they occur when they do.

CONTINUOUS VERSUS PULSATILE CONTROL

In the nervous system, action potentials last between 0.25 and 1.0 msec. Slower nervous events last between 4 and 5 msec, sometimes longer. Synaptic transmission takes between 2.0 and 500 msec. None of these times approaches the scale of diurnal, lunar or seasonal rhythms. Obviously, if the nervous system were to control such rhythms, it would have to act repeatedly--just one action potential is unlikely to do the job.

Table	
Hormone	**Half-life**
Thyroxine	6 days
Cortisol	0.07 days
Testosterone	0.04 days
Aldosterone	0.016 days
Growth hormone	0.017 days
Insulin	0.006 days

The situation is slightly better for the endocrine system. The half- lives of a number of hormones are shown in Table. With the exception of thyroxine, none of these hormones has a half-life longer than a day. The release of luteinzing hormone (LH) is known to be pulsatile, each pulse lasting about 30 minutes.

Some actions that must be controlled by the body are very brief, *e.g.,* many movements. Some appear to be very long, *e.g.,* growth. How can we get what appears to be continuous control with what must be a pulstile controller? Let's take an example. Muscle contractions can last for part of a second, a minute or longer, but the controller, the action potential in the motoneuron does not last more than 2 msec.

The answer lies in the time-constant, *i.e.*, the duration of action, of the controlled system. In our example, the controlled system is the muscle. Muscle contraction is a mechanical event, which is triggered by the action potential, an electrical event. The mechanical event is very slow (requires 250 or more msec for a single response) compared to the electrical event. The effect of two action potentials lasts even longer. If several action potentials follow each other closely in time down the axon, then the contraction of the muscle will be continuous and maximal for that muscle. In this way, a pulsatile controller can produce seemingly continuous control.

Chapter 4

REGULATORY CONTROL ANALYSIS

REGULATORY CONTROL IS THE FOUNDATION FOR ADVANCED PROCESS CONTROL

A dictionary definition of optimization might read: 'A strategy giving the best result obtainable under a given set of conditions.' To the practicing control engineer, optimization usually means a highly theoretical exercise, which is not really relevant in the real world, where pipes leak, sensors plug, pumps cavitate, and valves stick.

Optimization in an operating plant requires integrating process know-how to maximize productivity, including paying attention to field devices, control strategy suitability, controller tuning, and wise use of basic regulatory control (BRC)-cascade, feedback, feedforward, ratio, lead/lag, *etc.*-at the lowest possible level of implementation. It is upon this foundation that effective advanced process control (APC) can be constructed.

Maximizing return on investment (ROI) is always an important goal of plant management. However, purchasing control equipment, transmitters, and a digital control system loaded with the latest advanced control software does not guarantee good control. If not managed, understood, and optimized to match control objectives for which these purchases were made, the ROI is poor.

Function of Control Systems

Process control systems convert raw materials and energy into usable products through an intricate series of processing steps and control techniques. By themselves, plant processes do not produce levels, flows, or temperatures and the introduction of such variables represent-operating constraints designed to ensure productivity, efficiency, and/or quality product is produced.

For a plant to produce high quality product at the least costs requires a structure of people, systems, and facilities integrated together and built on a foundation of management commitment.

System Optimization Architecture Diagram

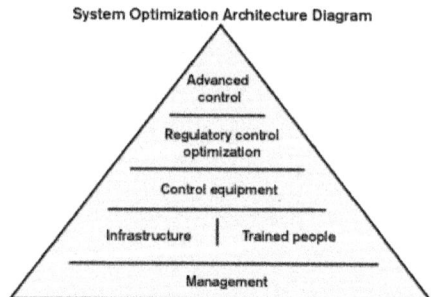

Among the most significant management challenges are cost containment, product quality, conforming to government regulations, and achievement of production schedule targets. Managers who recognize and embrace the value robust process control contributes toward overall operating results gain competitive advantage for their company. However, many managers have been 'burned' by past promises of improved process performance from control system investments that never materialized. Too frequently, the result has been underutilized control systems.

Gaining maximum advantage from the available capability of currently installed control systems requires an organizational philosophy that empowers, supports, and organizes in a way that achieves the desired results. Simply purchasing advanced hardware and software does not ensure success; without management's commitment any system optimization program is doomed to failure. Success requires effective management of the people, facilities, and systems working together to achieve established goals.

People and Systems

For any optimization process to be successful, an infrastructure of systems must be in place that provide the training and tools so qualified people can successfully carry out activities that lead to meeting defined goals and objectives. The introduction of high technology equipment-digital control systems, graphic monitors, smart transmitters, and artificial intelligence-into the work place do not reduce, but increase the requirement for training personnel.

Effectively optimizing regulatory functions of a process control system requires knowledge of the static and dynamic factors that exist between, and among, loops.

Control theory, as presented in university courses, is typically abstract mathematics. Once out of school, control engineers quickly forget what was learned and process control is practiced on an *ad hoc* basis without reference to the theory that governs the behavior of dynamic systems.

During the past sixteen years, Techmation (Scottsdale, Ariz.) consultants have worked in over 2,700 operating plants, testing, analyzing, and improving the dynamic operational characteristics of tens of thousands of control loops.

During this time procedures and data analysis techniques have been developed and refined into a knowledge base that bridges the gap between theory and real-world practice.

Because process operations are dynamic, it's important that after consultants complete and leave, customer employees can maintain the control system at the level of optimization achieved.

To ensure a satisfactory level of customer employee proficiency, consultants mix classroom theory with practical experience to form a process audit and improvement team.

Following classroom instruction on how to efficiently and effectively use a personal computer to test and analyze process data, the audit team, under the supervision of the consultant, apply a methodology of identifying, auditing, testing, analyzing, and optimizing to establish a solid foundation for APC. Techmation's methodology follows.

- Identification consists of designing and conducting tests that identify critical measurements and control loops.

- Auditing uses test data to determine the accuracy and capability of all measurements and loops. The team corrects problems discovered during the audit, such as calibration and valve errors. Critical loops are tested to determine installed characteristic. When installed characteristic are found to be highly non-linear, especially in the normal operating range, the problem is corrected.

Where required, controller algorithms are re-programmed and re-tuned to meet control requirements. If audit testing indicates installed control strategies do not meet objectives, the audit team designs, implements, tests, and documents new strategies that will meet defined control objectives. When completed, the audit report provides a 'capability signature' of the process personnel can use to maintain system integrity

- Analyzing and optimizing consists of using new data, collected following the audit phase, to identify where changes or new control strategies could improve unit control. During this phase, the emphasis is on identifying unit optimization strategies and system configuration implementation methods.

By following an identifying, auditing, testing, analyzing, and optimizing methodology and mixing formal classroom training with practical experience, under the guidance of a consultant, users can optimize regulatory control systems and ensure they stay optimized.

Process Control Equipment, Strategies

A poorly designed control strategy, inaccurate measurements, and poorly functioning control valves will result in less than optimum control.

Control system testing typically documents a number of equipment related problems that need to be fixed before the regulatory control systems dynamic operation can be optimized. Attempting to mask equipment problems by adjusting the PID (proportional/integral/derivative) and filter settings in controllers is often the response where issues of training, time constraints, and management commitment are manifest. In fact, as many as 50% of loops need some maintenance or configuration changes before the loop can be tuned to provide minimum variance regulatory control.

Regulatory control consists of two types of control functions that can be combined in an almost infinite number of configurations. The two types of regulatory control are: feedback and feedforward.

Feedback control can be configured for cascade, selective, ratio, and any number of other types of control schemes. All feedback control implementations have one thing in common. The controlled process variable measurement is compared to a reference, called the setpoint, and the deviation results in a corrective action by the controller. In short, feedback control only allows upsets to be corrected *after* they are detected. Tuning regulatory controls requires knowledge of the transfer function that describes the dynamic relationship between a change in the controller output and the response of the variable being controlled. Simplified first order models, rule of thumb procedures, and simplified rule based self-tuning controllers can, not in many cases, provide the best regulatory control solution for many of the dynamically complex processes found in control applications.

Feedforward control measures a load disturbance and introduces a dynamically compensated corrective action before the load disturbance affects the controlled variable. Tuning of feedforward control loops requires knowledge of the process transfer function in addition to the load upset transfer function. The only way to accurately obtain the necessary knowledge of the system dynamics is to test the installed system under operating conditions to obtain time domain data that accurately represents the process response. Time domain data must be accurately transformed into the frequency domain to obtain the best control solutions.

The function of the regulatory portion of the control system is to reduce variability in the face of changing conditions. Without an effective regulatory control system, each successive unit operation can introduce variation that can accumulate throughout the process and is reflected in the final product quality and overall cost of production. To produce a uniform product that consistently meets customer demands at the lowest cost, a regulatory control system must be in place to minimize variance throughout the processing cycle.

A modern process plant may have thousands of control loops. Advanced control systems cannot be operated as designed without a majority of these loops being in automatic control. The regulatory control loops provide four functions:

1. Allow the process to operate at a chosen target;
2. Minimize effects of load disturbances;

3. Reduce the effect of raw material variability; and

4. Provide for safe and efficient startup, operation, and shutdown of the process.

Hence, the regulatory control system's function is to maximize product uniformity under dynamic conditions.

Regulatory Controllers

PID is the most common feedback controller, has been in use for more than 60 years, yet is not covered by any implementation standard. In fact, no two digital control systems implement PID control in the same manner, understandably complicating tuning methods and calculations. For example, digital controller manufactures write software for PID controllers using the ideal, series, or parallel algorithm configurations. Because of PID implementation algorithm differences, loop-tuning parameters can be significantly different to accomplish the same task depending on the algorithm type implemented. Adding to the confusion, controller-tuning units can be expressed in different units and time domains such as:

- Proportional setting being in gain, percent proportional band, or throttling range;

- Integral setting in seconds per repeat, repeats per second, minutes per repeat, repeats per minute, or-in some cases-scan rate per repeat and repeats per scan rate; and

- Derivative terms expressed in seconds, minutes, or scan rate.

Muddying the waters still further, some manufactures allow the end-user to select the controller algorithm and tuning units. As if the preceding was not enough to create a sufficient tuning maze, add derivative filter constant settings, positional or velocity digital implementation of the PID, configurations for PID, PI-D, I-PD setpoint response, and PV filtering options, and it's easy to understand why 'simple' tuning can be confusing.

Conversely, over the years, numerous special linear and non-linear versions of the PID controller have been developed that provide better control of particular processes. For example, a non-linear PID controller algorithm is available that eliminates the stick-slip cycling found in as many as 30% of fast control loops.

Other special controller algorithms are used for averaging control in level systems, eliminating hysteresis cycling in integrating processes, preventing overshoot when filtering is used, and the conditional integral configuration for batch control to name just a few.

Understanding how to correctly implement these feedback controller algorithms is important to insure minimum variance control. Indeed, the crucial consideration in regulatory controllers is to understand the complex picture that exists with a trained eye toward the application of the appropriate techniques and solutions.

Testing Regulatory Loops

Attempts to achieve 'total product quality,' using SPC (statistical process control) and compliance with ISO quality standards, ignore control loop details such as:

- Correctly pairing of controlled and influential measurements;
- Hysteresis, stick-slip, and sizing of control valves; and
- Quality of the measurements, signal ailiasing problems, control algorithms, signal filtering, system configuration, and tuning of the regulatory control system.

Audit experience repeatedly indicates regulatory control systems are operational, but not providing optimum control. Findings indicate the typical regulatory control system contributes to as much as 50% of the non-uniformity of the final product. Testing of tens of thousands of unit operations with regulatory control systems applied consistently reveals the installed dynamics and loop tuning information, identifies equipment problems, installed characteristics of process loops, and relative gain of coupled loops.

Among the variety of problems identified, a few appear over and over in one of three general categories of: control valves, measurement, and control strategies.

Control Valve Problems

Stick-slip Cycling

Tests reveal as many as 30% of rotary and high friction globe valves exhibit a tendency to produce stick-slip cycle at steady state operating conditions when the controller is tuned based on installed loop dynamics.

The stick-slip characteristics in a control valve results in cycling at steady state and can produce excessive process variability in unit operations. Stick-slip is a result of an excessive ratio of static to dynamic friction in the control valve, pneumatic stiffness in the actuator, and the performance of the valve positioner.

The steady state cycling of the controller output and the process variable being controlled is a typical stick-slip cycle. A linear PID controller measures the error between the setpoint and the process variable and ramps the controller output at a rate that is a function of the controller tuning parameters to correct the error. After a small and typically slow ramp change in the controller output, the valve position 'jumps' to a new position and the flow overshoots the setpoint. The error being on the other side of the setpoint causes the controller to again ramp its output to correct the error. The valve again 'jumps' to a new position and the cycle is repeated.

High-friction valves typically require tuning parameters that are 10 to 20 times slower than required, based on the installed loop dynamics to eliminate stick-slip cycling.

Correcting this very common problem requires a non-linear PID algorithm that sets the controller integral value to a lesser (slower) value only when the error is very small in the range where stick-slip cycling occurs. This algorithm has been installed in thousands of loops on numerous different controller brands with a net result of rejecting fast load disturbances and 'smoothing' steady state operation.

Hysteresis

Loose linkages in the actuator or positioner mountings-combined with friction in the valve-cause hysteresis or deadband in pneumatic control valves. Loop analysis testing reveals the normalized magnitude of the hysteresis in each loop.

Small amounts of hysteresis can usually be tolerated in self-regulating loops but will result in continuous cycling at steady state in integrating loops such as level, large volume pneumatic pressure, and batch temperature loops.

Techniques to eliminate hysteresis cycling in integrating loops include:

- Fixing the control valve;
- Placing the integrating loop in cascade where the inner loop is a self-regulating process; or
- Implementing an error squared on integral control algorithm.

Installed Characteristic

No control loop has a completely linear installed characteristic. In most loops this non-linear installed characteristic can be easily handled using an appropriate gain margin in the calculation of the controller tuning parameters. In some instances the installed characteristic in the loop is so non-linear the loop must be made linear before the loop can be tuned for optimum closed loop response under all system load conditions. When this is the case software, such as Techmation's Protuner, is used to record the controller output in percent and the measured variable in percent. This data can be analyzed and used to determine the cam characteristic that will result in a linear response.

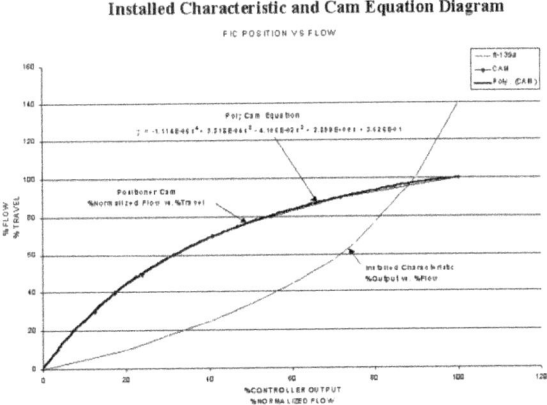

Installed Characteristic and Cam Equation Diagram

The illustration shows an equal percentage installed characteristic. Contrasted with linear characteristics, when the loop in the illustration is tuned with the valve at low-end travels, the loop will become unstable at high-end travels. Conversely, tuning the loop at high-end valve travel will cause sluggish performance at low-end valve travels. Even when the loop is tuned for stability under all loads, the non-linear installed characteristic will result in a varying closed-loop time constant. To operate properly, the APC closed loop model needs to be varied as a function of load.

The compensating cam is the mirror image of the normalized installed characteristic. The illustration indicates the cam characteristics graphically and as a polynomial equation, making it easy to implement in the controller output or as a digital cam in a digital (smart) positioner. Implementing the can will linearize the process and loop tuning will be effective under all load conditions with the added benefit of constant closed loop dynamic in support of APC modeling.

Other problems -Valve calibration and valve sizing are also common problems found.

A recent paper released by a major valve manufacture revealed as many as seventy percent of installed valves tested require zero and span calibration.

Incorrect zero and span of control valves can lead to various control problems not easily solved. For example, if the valve is calibrated to operate 0% to 100% travel from 10% to 80% controller outputs there is a potential integral windup problem. Windup can result in large and unexpected overshoot and slow startup of the affected variable.

Techmation's experience reveals as many as 30% of valves are oversized and 15% are undersized for the installed application. Oversized valves typically provide poor performance due to lack of rangibility; undersized valves can result in production bottlenecks.

Measurement Problems

A generally understood first law among process control engineers is, 'You can't control what you can't measure.'

In testing of regulatory control systems, impediments to accurately measure the variables being controlled is hindered by such things as:

- Lack of or improper setting of the required anti-aliasing filter constant in the transmitter;
- Excessive noise on the measurement signal;
- Improper use of or excessive signal filtering;
- Incorrect PID controller configurations when measurement filtering is present;
- Incorrect placement or mounting of measurement sensors; and
- Incorrect transmitter calibration and scaling.

Unless measurement issues are addressed, optimum regulatory control is impossible and APC cannot provide the increased system performance expected.

Control Strategy Problems

Experience indicates as many as 17% of regulatory unit operation control strategies are incorrectly implemented and must be redesigned. The following example illustrates the results of a typical control strategy redesign to control column pressure.

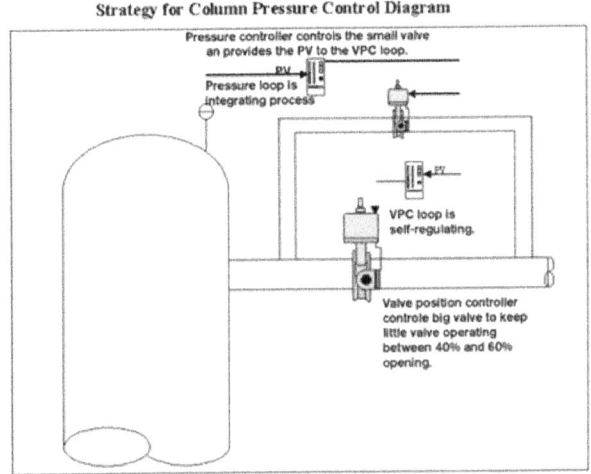

Strategy for Column Pressure Control Diagram

The original strategy used a single controller split ranged to control both valves. Testing the system revealed the pressure control loop was very non-linear using the original strategy. Because the process is a mass balance its dynamics are integrating. As shown in the 'Typical closed loop hysteresis cycle in level loop under PI control' diagram above, any hysteresis or deadband in the valve will result in continuous cycling at steady state. To retain the original control strategy, and make it effective, would require linearizing the loop in software, and replacing the large butterfly valve because it is not a precision control valve.

The new strategy retains the existing large butterfly valve and controls the column pressure using the small precision control valve. The VPC (valve position controller) controls the position of the large valve to keep the small valve in range. The VPC control loop is a self-regulating process where the hysteresis and dead-band does not result in cycling. Tuned correctly, the new control strategy provides accurate pressure control using existing valves with no cycling at steady state.

Regulatory Control Optimization Results

Poorly implemented control strategies, faulty equipment, improper setup, bad installation, lack of anti-aliasing filters, and Murphy's Law are the conditions that require on-site system analysis testing to tune the regulatory control system.

Over the last several years, a major company's engineers have used the Protuner System Analyzer and the test procedures learned from the consultants to optimize unit operations at a number of facilities, yet they still find room for improvement. This clarifies the fundamental need for conscientious and consistent attention to the optimization of the regulatory control system to ensure economic APC ROI is achieved.

The tuning of the single and interactive loops is ascertained based on the analysis of the actual installed dynamic transfer functions of the individual processes. Tuning parameters are determined based on pole cancellation with adequate gain and phase margins, and damping factor requirements as a function of the installed linearity of each loop.

Feedforward tuning is based upon the actual dynamic transfer function models of the process and load disturbance.

The net result has been better, safe, and more efficient operating plants for this customer.

Advanced Process Control

Advanced control systems are implemented to control the process, not as individual levels, flows, pressures, or temperatures, but rather as each variable relates to productivity or efficiency of the process. From an advanced control system viewpoint, the control system is not single loop controls but a multi-variable envelope viewed as a polygon, with each side representing the constraints of pressure, temperature, *etc.* Within the envelope, the process is continuously maximizing efficiency.

There are a large number of techniques employed that come under the general category of advanced process control. The most common, yet least discussed advanced control strategy, is operator knowledge and confidence the regulatory control system works. In many cases, no matter what control strategy is implemented, operators will set the individual process variable setpoints at 'safe,' though not necessarily optimum target setpoints. Until operators are confident the regulatory control system is capable of safe and reliable operation-at or near process limits-the operator safe factor will often result in undesirable process integrity.

Another advanced-or optimization control strategy-is the proper application of regulatory control that is designed and implemented to maximize efficiency of the operation.

Examples include the use of variable speed pumps and correlated control strategies to allow the control system to better follow demand. Additional widely discussed advanced control strategies include dynamic matrix control, fuzzy logic, and multi-variable control. Each has several things in common including the obvious goal to continuously adjust regulatory control to maximize the operational efficiency of the process. Therefore, the success of an advanced control system is directly impacted by how well the regulatory control system functions.

Years of experience reinforce there is no magic panacea for the optimizing of a control system. In some cases, an advanced control package is sold as the ultimate answer to reducing variability and improving efficiency without adequate consideration given to the underlying regulatory control systems operation. Expectations of this nature are too frequently disappointing and can give process control an unjustified 'black eye.'

Most regulatory control systems are de-tuned at startup for steady state operation to mask design and equipment related problems, or because of a lack of knowledge about the processes dynamics. De-tuning the regulatory control system avoids 'troublesome oscillation,' but almost always results in the need to constantly adjust regulatory loop setpoints to overcome upsets 'caused' by the APC.

Management insight and commitment, trained and motivated people, the proper tools, and plain hard work on the regulatory portion of the control system form the necessary foundation to successfully apply APC.

The knowledge gained during testing and analyzing regulatory loops provides exactly the knowledge needed when developing, applying, and tuning the APC models and establish the foundation for continuous process improvements.

FEED FORWARD (CONTROL)

Feed-forward, sometimes written feedforward, is a term describing an element or pathway within a control system which passes a controlling signal from a source in its external environment, often a command signal from an external operator, to a load elsewhere in its external environment. A control system which has only feed-forward behavior responds to its control signal in a pre-defined way without responding to how the load reacts; it is in contrast with a system that also has feedback, which adjusts the output to take account of how it affects the load, and how the load itself may vary unpredictably; the load is considered to belong to the external environment of the system.

In a feed-forward system, the control variable adjustment is not error-based. Instead it is based on knowledge about the process in the form of a mathematical model of the process and knowledge about or measurements of the process disturbances.

Some prerequisites are needed for control scheme to be reliable by pure feed-forward without feedback: the external command or controlling signal must be available, and the effect of the output of the system on the load should be known (that usually means that the load must be predictably unchanging with time). Sometimes pure feed-forward control without feedback is called 'ballistic', because once a control signal has been sent, it cannot be further adjusted; any corrective adjustment must be by way of a new control signal. In contrast 'cruise control' adjusts the output in response to the load that it encounters, by a feedback mechanism.

These systems could relate to control theory, physiology or computing.

Overview

With feed-forward control, the disturbances are measured and accounted for before they have time to affect the system. In the house example, a feed-forward system may measure the fact that the door is opened and automatically turn on the heater before the house can get too cold. The difficulty with feed-forward control is that the effect of the disturbances on the system must be accurately predicted, and there must not be any unmeasured disturbances. For instance, if a window was opened that was not being measured, the feed-forward-controlled thermostat might still let the house cool down.

The term has specific meaning within the field of CPU-based automatic control. The discipline of "feedforward control" as it relates to modern, CPU based automatic controls is widely discussed, but is seldom practiced due to the difficulty and expense of developing or providing for the mathematical model required to facilitate this type of control. Open-loop control and feedback control, often based on canned PID control algorithms, are much more widely used.

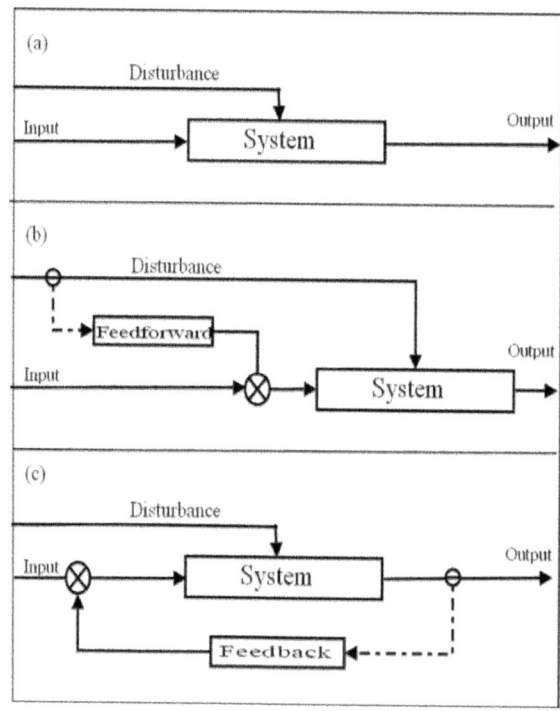

Fig. : The Three types of Control System (a) Open Loop (b) Feed-forward (c) Feedback (Closed Loop) Based on Hopgood (2002)

There are three types of control systems: open loop, feed-forward, and feedback. An example of a pure open loop control system is manual non-power-assisted steering of a motor car; the steering system does not have access to an auxiliary power source and does not respond to varying resistance to turning of the direction wheels; the driver must make that response without help from the steering system.

In comparison, power steering has access to a controlled auxiliary power source, which depends on the engine speed. When the steering wheel is turned, a valve is opened which allows fluid under pressure to turn the driving wheels. A sensor monitors that pressure so that the valve only opens enough to cause the correct pressure to reach the wheel turning mechanism. This is feed-forward control where the output of the system, the change in direction of travel of the vehicle, plays no part in the system.

If you include the driver in the system, then she does provide a feedback path by observing the direction of travel and compensating for errors by turning the steering wheel. In that case you have a feedback system, and the block labeled "System" in Figure(c) is a feed-forward system.

In other words, systems of different types can be nested, and the overall system regarded as a black-box.

Feedforward control is distinctly different from open loop control and teleoperator systems. Feedforward control requires a mathematical model of the plant (process and/or machine being controlled) and the plant's relationship to any inputs or feedback the system might receive. Neither open loop control nor teleoperator systems require the sophistication of a mathematical model of the physical system or plant being controlled. Control based on operator input without integral processing and interpretation through a mathematical model of the system is a teleoperator system and is not considered feedforward control.

History

Historically, the use of the term "feedforward" is found in works by D. M. MacKay as early as 1956. While MacKay's work is in the field of biological control theory, he speaks only of feedforward systems. MacKay does not mention "Feedforward Control" or allude to the discipline of "Feedforward Controls." MacKay and other early writers who use the term "feedforward" are generally writing about theories of how human or animal brains work.

The discipline of "feedforward controls" was largely developed by professors and graduate students at Georgia Tech, MIT, Stanford and Carnegie Mellon. Feedforward is not typically hyphenated in scholarly publications. Meckl and Seering of MIT and Book and Dickerson of Georgia Tech began the development of the concepts of Feedforward Control in the mid 1970s.

Benefits

The benefits of feedforward control are significant and can often justify the extra cost, time and effort required to implement the technology. Control accuracy can often be improved by as much an order of magnitude if the mathematical model is of sufficient quality and implementation of the feedforward control law is well thought out. Energy consumption by the feedforward control system and its driver is typically substantially lower than with other controls.

Stability is enhanced such that the controlled device can be built of lower cost, lighter weight, springier materials while still being highly accurate and able to operate at high speeds. Other benefits of feedforward control include reduced wear and tear on equipment, lower maintenance costs, higher reliability and a substantial reduction in hysteresis. Feedforward control is often combined with feedback control to optimize performance.

Model

The mathematical model of the plant (machine, process or organism) used by the feedforward control system may be created and input by a control engineer or it may be learned by the control system. Control systems capable of learning and/or adapting their mathematical model have become more practical as microprocessor speeds have increased. The discipline of modern feedforward control was itself made possible by the invention of microprocessors.

Feedforward control requires integration of the mathematical model into the control algorithm such that it is used to determine the control actions based on what is known about the state of the system being controlled. In the case of control for a lightweight, flexible robotic arm, this could be as simple as compensating between when the robot arm is carrying a payload and when it is not.

The target joint angles are adjusted to place the payload in the desired position based on knowing the deflections in the arm from the mathematical model's interpretation of the disturbance caused by the payload. Systems that plan actions and then pass the plan to a different system for execution do not satisfy the above definition of feedforward control. Unless the system includes a means to detect a disturbance or receive an input and process that input through the mathematical model to determine the required modification to the control action, it is not true feedforward control.

Open System

In systems theory, an open system is a feed forward system that does not have any feedback loop to control its output. In contrast, a closed system uses on a feedback loop to control the operation of the system. In an open system, the output of the system is not fed back into the input to the system for control or operation.

Applications

Physiological Feed-forward System

In physiology, feed-forward control is exemplified by the normal anticipatory regulation of heartbeat in advance of actual physical exertion. Feed-forward control can be likened to learned anticipatory responses to known cues. Feedback regulation of the heartbeat provides further adaptiveness to the running eventualities of physical exertion.

Feedforward systems are also found in biological control by human and animal brains. Even in the case of biological feedforward systems, such as in the human brain, knowledge or a mental model of the plant (body) can be considered to be mathematical as the model is characterized by limits, rhythms, mechanics and patterns.

A pure feed-forward system is distinct from a homeostatic control system, which has the function of keeping the internal environment of the body steady or constant or in a prolonged steady state of readiness, and relies mainly on feedback, indeed on negative feedback, in addition to the feedforward elements of the system.

Gene Regulation and Feed-forward

The cross regulation of genes can be represented by a graph, where genes are the nodes and one node is linked to another if the former is a transcription factor for the latter. A motif which predominantly appears in all known networks (E. coli, Yeast,...) is A activates B, A and B activate C. This motif has been shown to be a feed forward system, detecting non-temporary change of environment. This feed forward control theme is commonly observed in hematopoietic cell lineage development, where irreversible commitments are made.

Feed-forward Systems in Computing

In computing, feed-forward normally refers to a perceptron network in which the outputs from all neurons go to following but not preceding layers, so there are no feedback loops. The connections are set up during a training phase, which in effect is when the system is a feedback system.

Long Distance Telephony

In the early 1970s, intercity coaxial transmission systems, including L-carrier, used feed-forward amplifiers to diminish linear distortion. This more complex method allowed wider bandwidth than earlier feedback systems. Optical fiber, however, made such systems obsolete before many were built.

Automation and Machine Control

Feedforward control is a discipline within the field of automatic controls used in automation.

TROUBLESHOOTING AND SOLVING POOR
CONTROL LOOP PERFORMANCE

Most power plants have a few control loops that never seem to control satisfactorily. These loops may oscillate for seemingly no reason or deviate far from their set points during load ramps and disturbances, often overshooting their set points afterward. Tuners can spend many hours trying to tune these challenging control loops, but the loops often remain problematic.

Poor control performance makes itself evident primarily when a controlled variable (process variable) deviates excessively from its target value (set point). Excessive and/or unnecessary deviations from set point can result from oscillations, sluggish control loop response, and excessive measurement noise. Each of these can have several causes.

Table: Common problems and possible causes of poor control performance.

Problem	Possible causes
Oscillations	Valve/damper stiction
	Controller tuning
	Cyclical interaction
	Deadband
	Process issues
Sluggish response	Controller tuning
	Excessive filtering
	Improper disturbance handling
	Deadband
Noise	Incorrect measurement technology
	Improper installation of measurement device
	Turbulence or volatility
	Insufficient filtering

Oscillations are periodic deviations from set point, frequently intermixed with some random behavior. Sluggish response is a general slowness in the loop's ability to recover from disturbances or follow set point changes. Noise is random movement in the process variable.

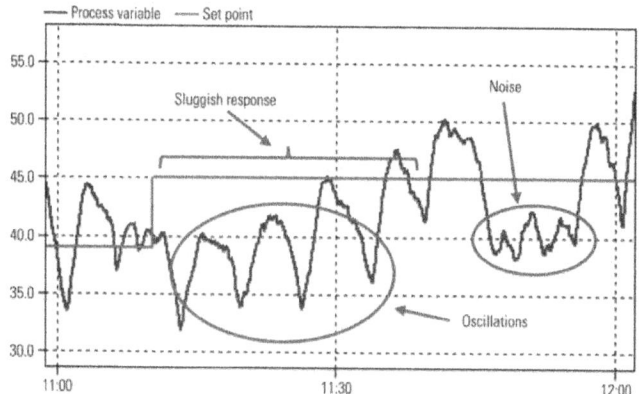

1. Random moves. Oscillations are cyclical deviations from the set point and often contain a component of randomness. Other control response problems include sluggish response and noise.

Troubleshooting Oscillations

Oscillations can originate from within the control loop or may be caused by external factors. Oscillations may also be a result of a cyclical interaction between two or more control loops. To narrow down the possible causes of the oscillation, the controller should be put in manual mode to see if the oscillation stops. If the oscillation persists when the controller is in manual mode, the oscillation originates from outside the loop. Oscillation analysis of interacting loops and equipment should then be done to find the root cause. For example, steam generator furnace pressure can oscillate because of a partially blocked rotating air heater.

If a control loop's set point oscillates, it will most likely cause the entire loop to oscillate. The cause of oscillations should be traced back to the origin of the loop's set point. For example, if a desuperheater outlet temperature control loop oscillates because its set point is oscillating, the problem likely lies with the main steam temperature controller producing the set point. To verify this, put the main steam temperature controller in manual and see if the desuperheater outlet temperature controller stops oscillating. If it does, troubleshooting should be focused on the main steam temperature controller.

There are three principal factors that trigger oscillation of boiler control loops: interaction between loops, stiction, and deadband.

Cyclical Interactions

Tightly coupled boiler control loops with similar dynamics can begin to oscillate against each other. A good example of this is feedwater flow controlled with a variable-speed pump and differential pressure controlled with a feedwater control valve. The pump and valve each affect both the flow and the pressure, and a cyclical interaction between the flow and pressure control loops can easily occur. Cyclical interaction can be aggravated by aggressive tuning, for example when using quarter-amplitude-damping tuning methods, such as the Ziegler-Nichols or Cohen-Coon tuning rules without reducing the calculated controller gain by at least 50%.

The simplest technique for solving problems with cyclical interactions is to tune one of the interacting control loops to produce an overdamped response. The IMC (Lambda) tuning method can be used to obtain very stable control loops and has been proven to help settle down cyclical interactions.

Because a boiler consists of many interactive sub-processes, one oscillating loop can cause several other loops on the boiler to oscillate with it. The loops will all oscillate with the same period of oscillation. Historical trends or process analysis software can be used to identify all the loops on a boiler oscillating with the same period. The problem loop can be isolated through knowledge of the boiler and its interactions, by looking at phase shifts between oscillations, or by placing likely culprit loops in manual one at a time.

For example, an oscillating reheat steam temperature control loop can cause oscillations in the generator load, throttle pressure, fuel, and drum level controllers. If the reheat temperature controller is put in manual control mode, the oscillations in all loops will cease. The reheat temperature control loop should be analyzed further. Note that this scenario is different from cyclical interactions in which two or more control loops interact directly with each other in a cyclical fashion. In the case of a cyclical interaction, any one of the participating control loops put in manual will cause all loops to stop oscillating.

Stiction

One of the leading causes of oscillations is stiction, short for static friction. When there is stiction, the final control element (for example, the control valve or damper) is mechanically sticky. The combination of a controller's integral action and a sticky control element causes an oscillation called a stick-slip cycle. When loop oscillations are caused by stiction, the controller output's trend often resembles a saw-tooth or triangular wave, while the process variable may look like a square wave or an irregular sine wave.

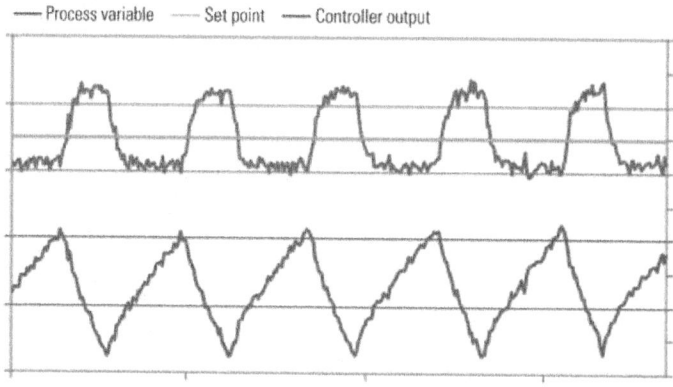

Fig. : Example of a stick-slip cycle.

Stiction can be detected by placing the controller in manual mode and making small changes (typically 0.5%) in controller output followed by monitoring the process variable for a resulting change. If the control valve seems to accumulate a few of the controller output changes before the process variable responds to them all at once, the control valve has stiction.

Deadband

Deadband in a final control element is a mechanical defect causing a dead zone through which the controller output must change before the control element responds after a change in travel direction. Many control systems also provide an adjustable deadband around a controller's set point. The latter is sometimes used to prevent the controller from reacting to process noise.

If an integrating control loop, such as mill or feedwater heater level control, drives a valve or damper directly, and the valve has deadband, the loop will oscillate. An integrating control loop will also oscillate if the controller has a deadband feature that is set wider than the measurement noise band.

Both kinds of deadband also reduce the effectiveness with which a controller can eliminate disturbances. Every time the direction of a disturbance changes, the process variable has to traverse the internal deadband before the controller begins responding, and the controller output has to traverse the controller's final control element's deadband before the latter actually begins moving. This causes the control loop to perform sluggishly. The speed of tuning should never be increased to compensate for deadband, because the aggressive tuning remains active after the deadband has been traversed, which can cause stability problems. Deadband in final control elements can severely affect the accuracy of controller tuning, often resulting in overly aggressive tuning.

Deadband can be detected with a simple process test consisting of two controller output steps in one direction and one step in the opposite direction with the controller in manual mode. The second and last steps should be the same size to make it easy to visually detect deadband from process trends. If the process variable does not reach the same level after the first and third steps, it indicates the presence of deadband. Final control elements with deadband must be repaired for optimum control.

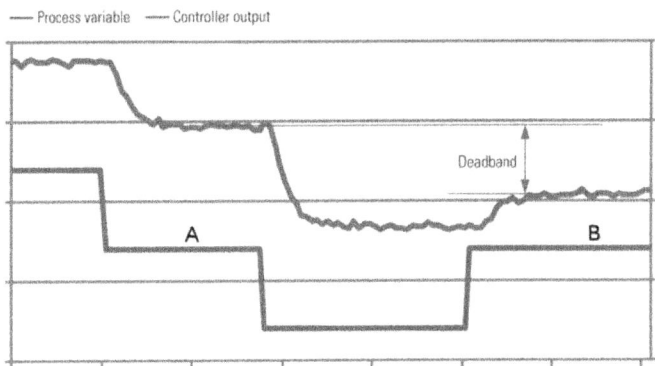

3. A deadband test. At time-slice A and B the controller output is at the same level. Because of deadband, the process variable does not return to the same level.

Tuning Control Loops

Incorrectly tuned control loops can respond sluggishly, or they can overshoot and oscillate significantly. Controller tuning is often done in a trial-and-error way, but this is not nearly as effective as using tuning rules or software. Controller tuning should start with step tests on the process to determine the dominant process characteristics: process gain, dead time, and time constant. Four or more step tests should be done, the results compared, outliers eliminated, and the average used for tuning. If process characteristics change with unit load or operating conditions, gain scheduling should be used.

Note that using the IMC tuning method results in stable control loops, but it often leaves a sluggish response to disturbances, especially on slow loops like main steam temperature. Cohen-Coon tuning provides faster disturbance rejection, even with the detuning recommended earlier. Quarter-amplitude-damping tuning methods should not be used because they result in marginally stable loops and oscillatory plants.

When tuning control loops, it is important to understand that any control loop can hypothetically be tuned for a slower response, but that is not always the case for a faster response. As a controller is tuned for a progressively faster response, the control loop's stability is compromised, and after some point it begins oscillating. A control loop's settling time is therefore limited to a certain minimum. No control loop recovering from a disturbance or responding to set point change can settle out faster than this minimum settling time.

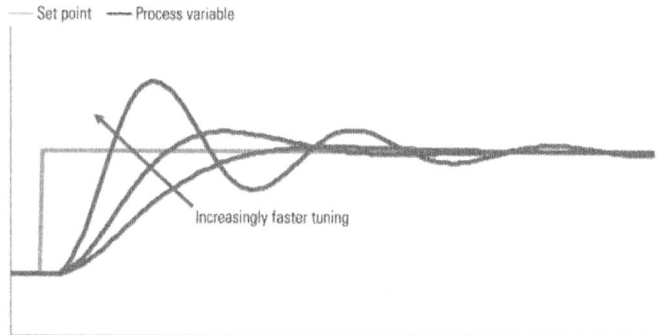

4. Tune for response, not speed. A control loop becomes less stable and more oscillatory with increasing speed of response.

The minimum settling time is virtually equal to the natural period of the process being controlled. It will take a control loop at least as long as the length of its process's natural period to settle out after a disturbance or set point change. The natural period of a process is determined mostly by its effective dead time, but the amount of lag (time-constant) also plays a role.

Table: Natural periods of different process types.

Process type	Natural period or minimum settling time
Dead time >> time constant	2 × dead time
Dead time ≈ time constant	3 × dead time
Dead time << time constant	4 × dead time
Integrating process	4 × dead time

Control loops cannot effectively respond to disturbances or set point changes shorter in duration than the loop's natural period (or settling time). Although correct tuning methods can go a long way in minimizing the effects of disturbances, if a disturbance occurs faster than the control loop can respond, there is very little the controller can do to reduce its amplitude. Then cascade and feedforward control should be considered.

Noise in Control Loops

Because noise-based deviations occur faster than the loop's settling time, it is impossible for the controller to eliminate the noise or even reduce its amplitude. Furnace pressure, with its rapid deviations because of turbulence and unstable combustion, is a good example of this. A controller responding aggressively to noise will likely increase variability. It will also cause unnecessary wear on the final control element, and it may disturb downstream control loops.

Assuming the correct measurement technology is being used, and the measurement device is properly installed, the amplitude of noise can often be reduced through filtering the process variable with a first-order lag filter. The filtering should be applied inside the control system (not in a replacement transmitter), either as a setting on the analog input card or in the controller itself. It is important to note that a filter increases the effective dead time in a control loop and therefore increases its minimum settling time. A filter can also hide real process problems and unsafe conditions. Therefore, filtering should be applied only when needed, and then as little of it as possible.

Chapter 5

TUNING OF PID CONTROLLERS

PID CONTROLLER TUNING IN SIMULINK

Introduction of the PID Tuner

PID Tuner provides a fast and widely applicable single-loop PID tuning method for the Simulink® PID Controller blocks. With this method, you can tune PID controller parameters to achieve a robust design with the desired response time.

A typical design workflow with the PID Tuner involves the following tasks:

(1) Launch the PID Tuner. When launching, the software automatically computes a linear plant model from the Simulink model and designs an initial controller.

(2) Tune the controller in the PID Tuner by manually adjusting design criteria in two design modes. The tuner computes PID parameters that robustly stabilize the system.

(3) Export the parameters of the designed controller back to the PID Controller block and verify controller performance in Simulink.

Opening the Model

Open the engine speed control model with PID Controller block and take a few moments to explore it.

```
open_system('scdspeedctrlpidblock');
```

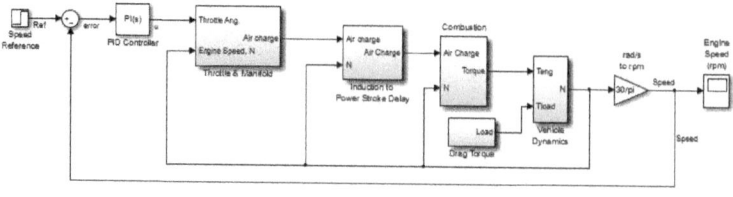

Design Overview

In this example, you design a PI controller in an engine speed control loop. The goal of the design is to track the reference signal from a Simulink step block scd-speedctrlpidblock/Speed Reference. The design requirement are:

- Settling time under 5 seconds
- Zero steady-state error to the step reference input.

In this example, you stabilize the feedback loop and achieve good reference tracking performance by designing the PI controller scdspeedctrl/PID Controller in the PID Tuner.

Opening the PID Tuner

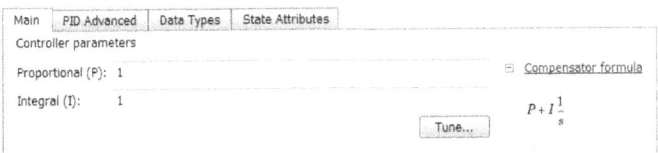

Initial PID Design

When the PID Tuner launches, the software computes a linearized plant model seen by the controller. The software automatically identifies the plant input and output, and uses the current operating point for the linearization. The plant can have any order and can have time delays.

The PID Tuner computes an initial PI controller to achieve a reasonable tradeoff between performance and robustness. By default, step reference tracking performance displays in the plot.

The following figure shows the PID Tuner dialog with the initial design:

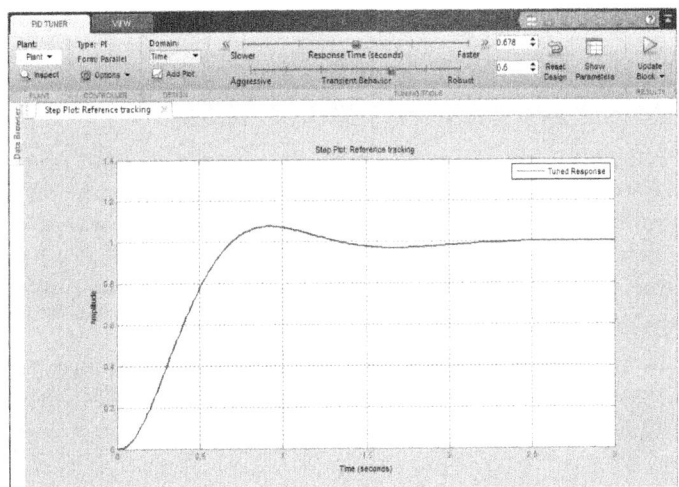

Displaying PID Parameters

Click Show parameters to view controller parameters P and I, and a set of performance and robustness measurements. In this example, the initial PI controller design gives a settling time of 2 seconds, which meets the requirement.

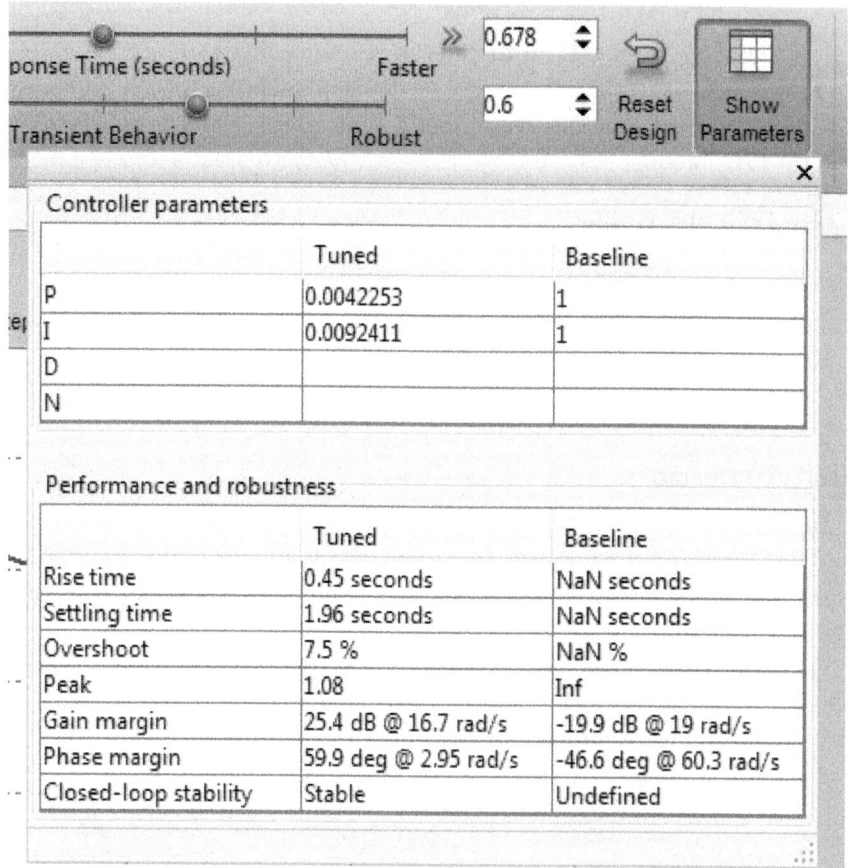

Controller parameters		
	Tuned	Baseline
P	0.0042253	1
I	0.0092411	1
D		
N		

Performance and robustness		
	Tuned	Baseline
Rise time	0.45 seconds	NaN seconds
Settling time	1.96 seconds	NaN seconds
Overshoot	7.5 %	NaN %
Peak	1.08	Inf
Gain margin	25.4 dB @ 16.7 rad/s	-19.9 dB @ 19 rad/s
Phase margin	59.9 deg @ 2.95 rad/s	-46.6 deg @ 60.3 rad/s
Closed-loop stability	Stable	Undefined

Adjusting PID Design in the PID Tuner

The overshoot of the reference tracking response is about 7.5 percent. Since we still have some room before reaching the settling time limit, you could reduce the overshoot by increasing the response time. Move the response time slider to the left to increase the closed loop response time. Notice that when you adjust response time, the response plot and the controller parameters and performance measurements update.

The following figure shows an adjusted PID design with an overshoot of zero and a settling time of 4 seconds. The designed controller effectively becomes an integral-only controller.

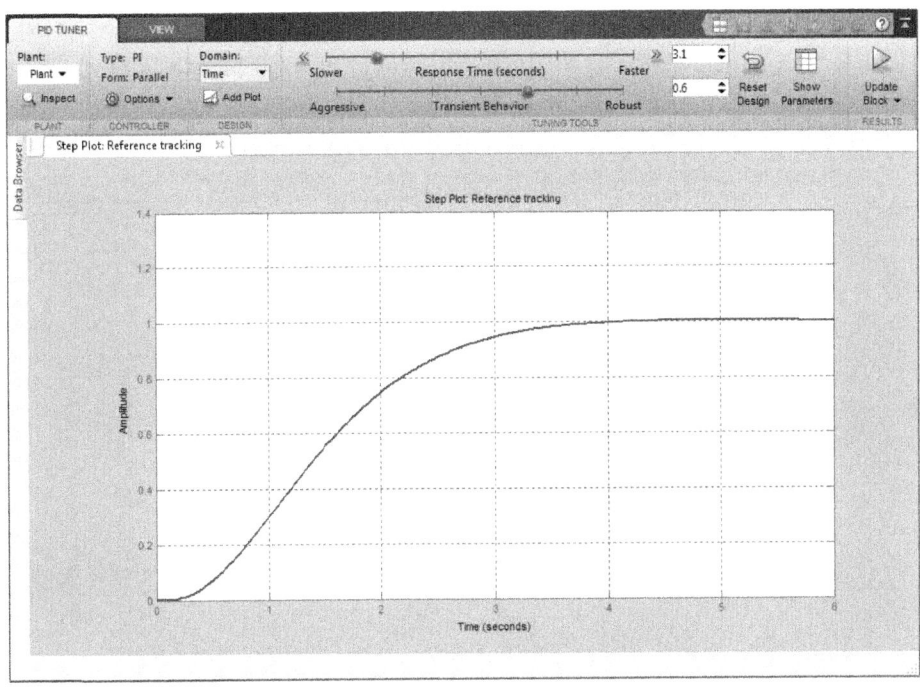

Controller parameters		
	Tuned	Baseline
P	0	1
I	0.0021263	1
D		
N		

Performance and robustness		
	Tuned	Baseline
Rise time	2.06 seconds	NaN seconds
Settling time	3.45 seconds	NaN seconds
Overshoot	0.401 %	NaN %
Peak	1	Inf
Gain margin	18.9 dB @ 3.27 rad/s	-19.9 dB @ 19 rad/s
Phase margin	69.3 deg @ 0.645 rad/s	-46.6 deg @ 60.3 rad/s
Closed-loop stability	Stable	Undefined

Completing PID Design with Performance Trade-Off

In order to achieve zero overshoot while reducing the settling time below 2 seconds, you need to take advantage of both sliders. You need to make control response faster to reduce the settling time and increase the robustness to reduce the overshoot. For example, you can reduce the response time from 3.4 to 1.5 seconds and increase robustness from 0.6 to 0.72.

The following figure shows the closed-loop response with these settings:

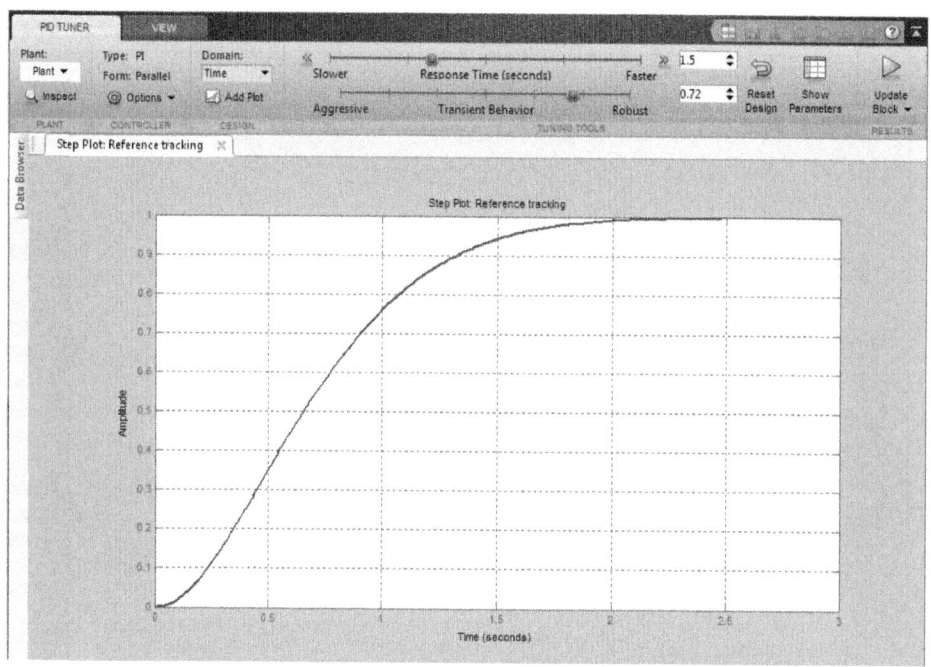

Controller parameters

	Tuned	Baseline
P	0.0014551	1
I	0.0043791	1
D		
N		

Performance and robustness

	Tuned	Baseline
Rise time	1.09 seconds	NaN seconds
Settling time	1.81 seconds	NaN seconds
Overshoot	0 %	NaN %
Peak	0.999	Inf
Gain margin	32.8 dB @ 15 rad/s	-19.9 dB @ 19 rad/s
Phase margin	72 deg @ 1.33 rad/s	-46.6 deg @ 60.3 rad/s
Closed-loop stability	Stable	Undefined

Writing the Tuned Parameters to PID Controller Block

After you are happy with the controller performance on the linear plant model, you can test the design on the nonlinear model. To do this, click Update Block in the PID Tuner. This action writes the parameters back to the PID Controller block in the Simulink model.

The following figure shows the updated PID Controller block dialog:

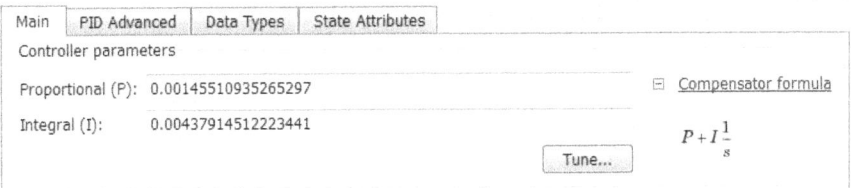

Completed Design

The following figure shows the response of the closed-loop system:

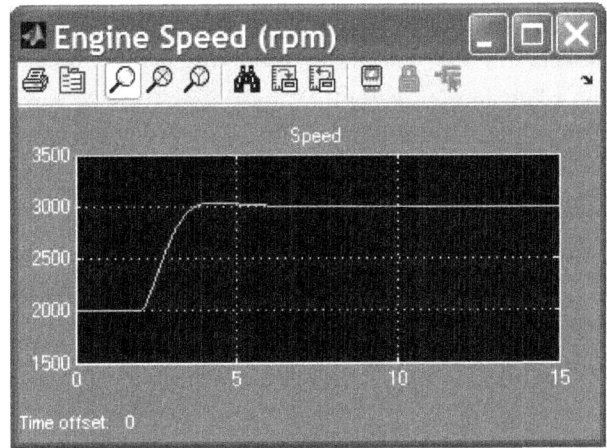

The response shows that the new controller meets all the design requirements.

You can also use the SISO Compensator Design Tool to design the PID Controller block. When the PID Controller block belongs to a multi-loop design task.

OPEN THE PID TUNER

Prerequisites for PID Tuning

Before you can use the PID Tuner, you must:

- Create a Simulink model containing a PID Controller or PID Controller (2DOF) block. Your model can have one or more PID blocks, but you can only tune one PID block at a time.

    o   If you are tuning a multi-loop control system with coupling between the loops, consider using other Simulink Control Design™ tools instead of the PID Tuner. The PID Controller blocks support vector signals. However, using the PID Tuner requires scalar signals at the block inputs. That is, the PID block must represent a single PID controller.

       Your plant (all blocks in the control loop other than the controller) can be linear or nonlinear. The plant can also be of any order, and have any time delays.

- Configure the PID block settings, such as controller type, controller form, time domain, sample time.

Opening the Tuner

To open the PID Tuner and view the initial compensator design:

1. Open the Simulink model by typing the model name at the MATLAB command prompt.
2. Double-click the PID Controller block to open the block dialog box.
3. In the block dialog box, click Tune to launch the PID Tuner.

When you open the PID Tuner, the following actions occur:

Note: If the plant model in the PID loop linearizes to zero, the PID Tuner provides the **Obtain plant model** dialog box. This dialog box allows you to obtain a new plant model by either:

    o Linearizing at a different operating point.

- The PID Tuner computes an initial compensator design for the linearized plant model using the algorithm described in PID Tuning Algorithm.

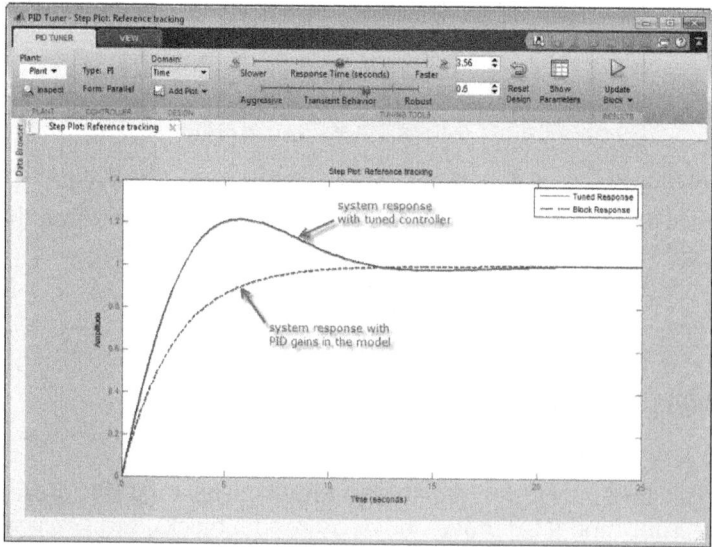

Tip After the tuner opens, you can close the PID Controller block dialog box.

- The PID Tuner displays the closed-loop step reference tracking response for the initial compensator design in the PID Tuner dialog box. For comparison, the display also includes the closed-loop response for the gains specified in the PID Controller block, if that closed loop is stable, as shown in the following figure.

ANALYZE DESIGN IN PID TUNER

Plot System Responses

To determine whether the compensator design meets your requirements, you can analyze the system response using the response plots. In the PID Tuner tab, select a response plot from the Add Plot menu. The Add Plot menu also lets you choose from several step plots (time-domain response) or Bode plots (frequency-domain response).

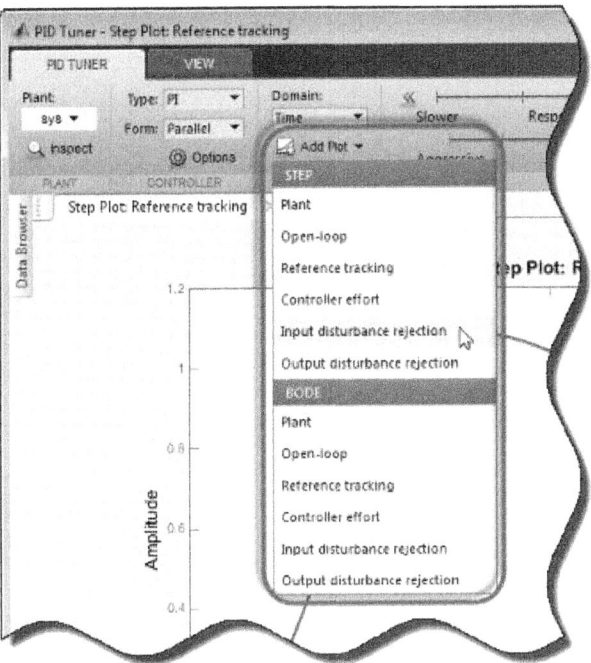

For 1-DOF PID controller types such as PI, PIDF, and PDF, PID Tuner computes system responses based upon the following single-loop control architecture:

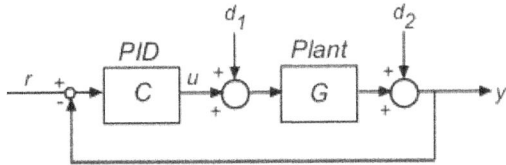

For 2-DOF PID controller types such as PI2, PIDF2, and I-PD, PID Tuner computes responses based upon the following architecture:

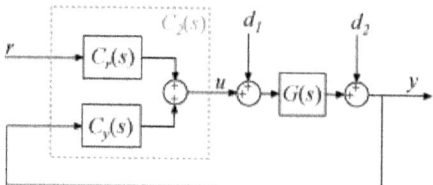

The system responses are based on the decomposition of the 2-DOF PID controller, C_2, into a setpoint component Cr and a feedback component Cy, as described in Two-Degree-of-Freedom PID Controllers.

The following table summarizes the available responses for analysis plots in PID Tuner.

Response	Plotted System (1-DOF)	Plotted System (2-DOF)	Description
Plant	G	G	Shows the plant response. Use to examine plant dynamics.
Open-loop	GC	$-GC_y$	Shows response of the open-loop controller-plant system. Use for frequency-domain design. Use when your design specifications include robustness criteria such as open-loop gain margin and phase margin.
Reference tracking	GC1+GC (from r to y)	GCr1−GCy (from r to y)	Shows the closed-loop system response to a step change in setpoint. Use when your design specifications include setpoint tracking.
Controller effort	C1+GC (from r to u)	Cr1−GCy (from r to u)	Shows the closed-loop controller output response to a step change in setpoint. Use when your design is limited by practical constraints, such as controller saturation.
Input disturbance rejection	G1+GC (from d_1 to y)	G1−GCy (from d_1 to y)	Shows the closed-loop system response to load disturbance (a step disturbance at the plant input). Use when your design specifications include input disturbance rejection.
Output disturbance rejection	11+GC (from d_2 to y)	11−GCy (from d_2 to y)	Shows the closed-loop system response to a step disturbance at plant output. Use when you want to analyze sensitivity to measurement noise.

Compare Tuned Response to Block Response

By default, PID Tuner plots system responses using both:

- The PID coefficient values in the PID Controller block in the Simulink model (Block response).
- The PID coefficient values of the current PID Tuner design (Tuned response).

As you adjust the current PID Tuner design, such as by moving the sliders, the Tuned response plots change, while the Block response plots do not.

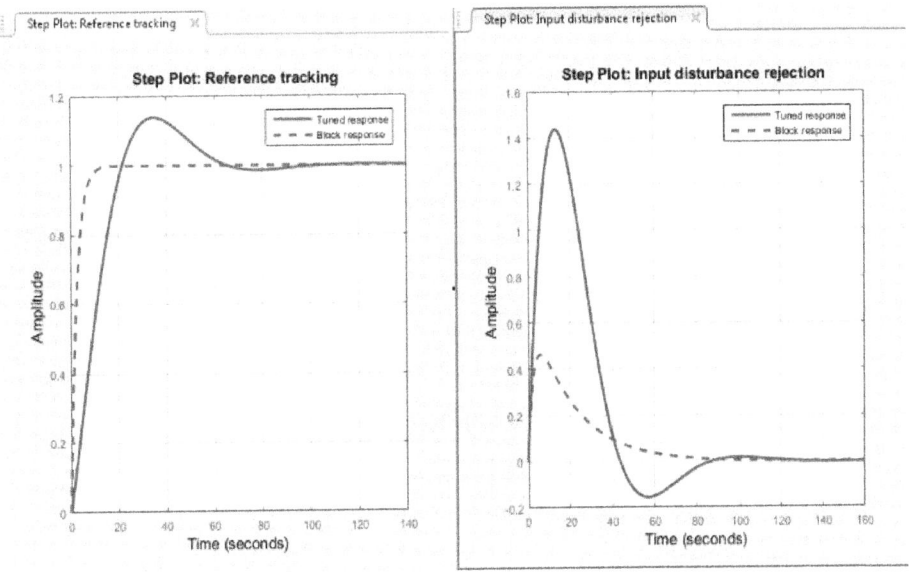

To write the current PID Tuner design to the Simulink model. When you do so, the current Tuned response becomes the Block response. Further adjustment of the current design creates a new Tuned response line.

View Numeric Values of System Characteristics

You can view the values for system characteristics, such as peak response and gain margin, either:

- Directly on the response plot — Use the right-click menu to add characteristics, which appear as blue markers. Then, left-click the marker to display the corresponding data panel.
- In the Performance and robustness table.

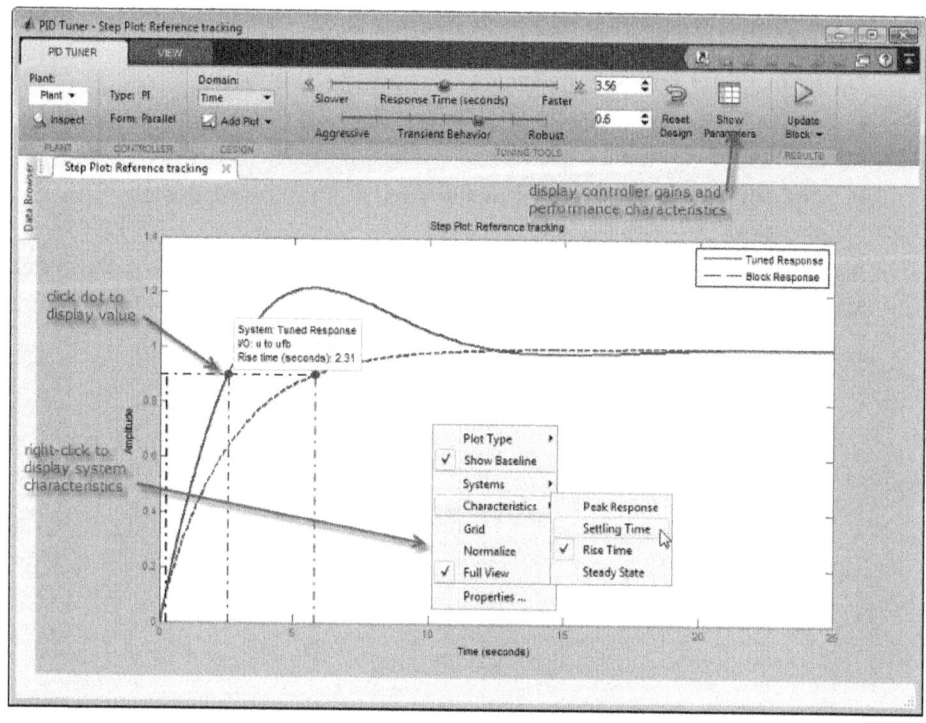

Export Plant or Controller to MATLAB Workspace

You can export the linearized plant model computed by PID Tuner to the MATLAB workspace for further analysis. To do so, click Update Block and select Export.

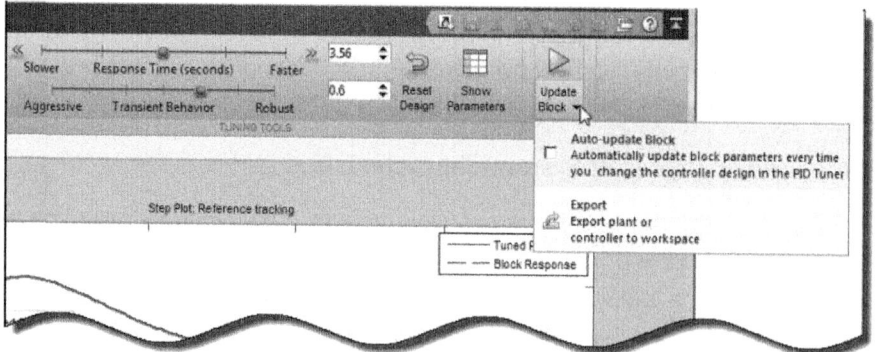

In the Export dialog box, check the models you want to export. Click OK to export the plant or controller to the MATLAB workspace as state-space (ss) model object or pid object, respectively.

Alternatively, you can export a model using the right-click menu in the Data Browser. To do so, click the Data Browser tab.

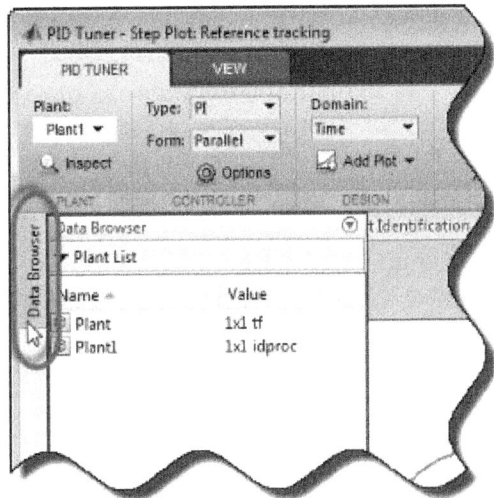

Then, right-click the model and select Export.

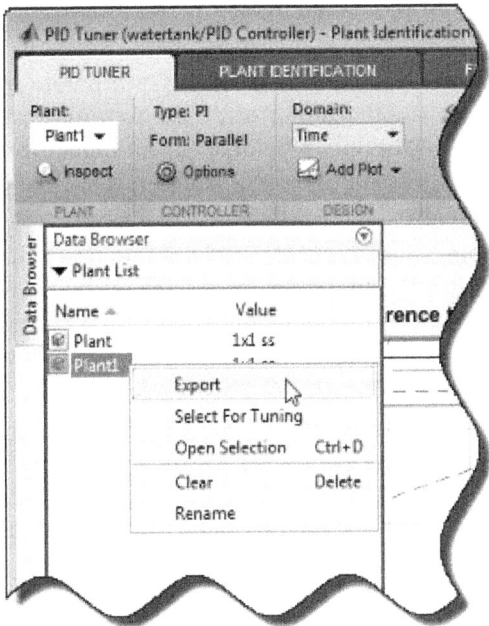

Refine the Design

If the response of the initial controller design does not meet your requirements, you can interactively adjust the design. The PID Tuner gives you two Domain options for refining the controller design:

- **Time domain (default)** — Use the Response Time slider to make the closed-loop response of the control system faster or slower. Use the Transient Behavior slider to make the controller more aggressive at disturbance rejection or more robust against plant uncertainty.

- **Frequency** — Use the Bandwidth slider to make the closed-loop response of the control system faster or slower (the response time is 2/ w_c, where w_c is the bandwidth). Use the Phase Margin slider to make the controller more aggressive at disturbance rejection or more robust against plant uncertainty.

Once you find a compensator design that meets your requirements, verify that it behaves in a similar way in the nonlinear Simulink model.

TUNING A PID (THREE-MODE) CONTROLLER

Controller Operation

There are three common types of Temperature/process controllers: ON/OFF, PROPORTIONAL, and PID (PROPORTIONAL INTEGRAL DERIVATIVE).

On/Off Control

An on-off controller is the simplest form of temperature control device. The output from the device is either on or off, with no middle state. An on/off controller will switch the output only when the temperature crosses the setpoint. For heating control, the output is on when the temperature is below the setpoint, and off above the setpoint.

Although capable of more complex control functions, the NEWPORT microprocessor based MICRO-INFINITY ® AUTOTUNE PID 1/16 DIN Controller can be operated as a simple On/Off Controller. The NEWPORT INFINITY ® series and INFINITY C ® series of highly accurate microprocessor based digital panel meters can all function as simple On/Off controllers.

With simple On/Off control, since the temperature crosses the setpoint to change the output state, the process temperature will be cycling continually, going from below setpoint to above, and back below. In cases where this cycling occurs rapidly, and to prevent damage to contactors and valves, an on-off differential, or "hysteresis," is added to the controller operations. This differential requires that the temperature exceed setpoint by a certain amount before the output will turn off or on again. On-off differential prevents the output from "chattering" or fast, continual switching if the temperature cycling above and below setpoint occur very rapidly.

"On-Off" is the most commonly used form of control, and for most applications it is perfectly adequate. It's used where a precise control is not necessary, in systems which cannot handle the energy being turned on and off frequently, and where the mass of the system is so great that temperatures change extremely slowly.

Backup alarms are typically controlled with "On-Off" relays. One special type of on-off control used for alarm is a limit controller. This controller uses a latching relay, which must be manually reset, and is used to shut down a process when a certain temperature is reached.

Proportional Control

Proportional control is designed to eliminate the cycling above and below the setpoints associated with On-Off control. A proportional controller decreases the average power being supplied to a heater for example, as the temperature approaches setpoint. This has the effect of slowing down the heater, so that it will not overshoot the setpoint, but will approach the setpoint and maintain a stable temperature.

This proportioning action can be accomplished by different methods. One method is with an analog control output such as a 4-20 mA output controlling a valve or motor for example. With this system, with a 4 mA signal from the controller, the valve would be fully closed, with 12 mA open halfway, and with 20 mA fully open.

Another method is "time proportioning" *i.e.* turning the output on and off for short intervals to vary the ratio of "on" time to "off" time to control the temperature or process.

With the analog output option, the NEWPORT INFINITY ® series and INFINITY C ® series of 1/8 DIN digital panel meters can function as proportional controllers. In addition, NEWPORT offers models of "INFINITY C" for thermocouple and RTD inputs featuring Time-Proportioning Control with its built in mechanical relays.

With proportional control, the proportioning action occurs within a "proportional band" around the setpoint temperature. Outside this band, the controller functions as an on-off unit, with the output either fully on (below the band) or fully off (above the band). However, within the band, the output is turned on and off in the ratio of the measurement difference from the setpoint. At the setpoint (the midpoint of the proportional band), the output on:off ratio is 1:1; that is, the on-time and off-time are equal. If the temperature is further from the setpoint, the on- and off-times vary in proportion to the temperature difference. If the temperature is below setpoint, the output will be on longer; if the temperature is too high, the output will be off longer.

The proportional band is usually expressed as a percent of full scale, or degrees. It may also be referred to as gain, which is the reciprocal of the band. Note, that in time proportioning control, full power is applied to the heater, but cycled on and off, so the average time is varied. In most units, the cycle time and/or proportional band are adjustable, so that the controller may be better matched to a particular process.

One of the advantages of proportional control is the simplicity of operation. However, the proportional controller will generally require the operator to manually "tune" the process, *i.e.* to make a small adjustment (manual reset) to bring the temperature to setpoint on initial startup, or if the process conditions change significantly.

Systems that are subject to wide temperature cycling need proportional control. Depending on the precision required, some processes may require full "PID" control.

PID (Proportional Integral Derivative)

Processes with long time lags and large maximum rate of rise (*e.g.*, a heat exchanger), require wide proportional bands to eliminate oscillation. The wide band can result in large offsets with changes in the load. To eliminate these offsets, automatic reset (integral) can be used. Derivative (rate) action can be used on processes with long time delays, to speed recovery after a process disturbance.

The most sophisticated form of discrete control available today combines PROPORTIONAL with INTEGRAL and DERIVATIVE or PID .

The NEWPORT MICRO-INFINITY® is a full function "Autotune" (or self-tuning) PID controller which combines proportional control with two additional adjustments, which help the unit automatically compensate to changes in the system. These adjustments, integral and derivative, are expressed in time-based units; they are also referred to by their reciprocals, RESET and RATE, respectively.

The proportional, integral and derivative terms must be individually adjusted or "tuned" to a particular system.

It provides the most accurate and stable control of the three controller types, and is best used in systems which have a relatively small mass, those which react quickly to changes in energy added to the process. It is recommended in systems where the load changes often, and the controller is expected to compensate automatically due to frequent changes in setpoint, the amount of energy available, or the mass to be controlled.

The "autotune" or self-tuning function means that the MICRO-INFINITY will automatically calculate the proper proportional band, rate and reset values for precise control.

Temperature Control

Tuning a PID (Three-Mode) Controller

Tuning a temperature controller involves setting the proportional, integral, and derivative values to get the best possible control for a particular process. If the controller does not include an autotune algorithm or the autotune algorithm does not provide adequate control for the particular application, the unit must then be tuned using a trial and error method.

The following is a tuning procedure for the NEWPORT® MICRO-INFINITY ® controller. It can be applied to other controllers as well. There are other tuning procedures which can also be used, but they all use a similar trial and error method. Note that if the controller uses a mechanical relay (rather than a solid state relay) a longer cycle time (10 seconds) is recommended when starting out.

The following definitions may be needed:

- **Cycle time** — Also known as duty cycle; the total length of time for the controller to complete one on/off cycle. Example: with a 20 second cycle time, an on time of 10 seconds and an off time of 10 seconds represents a 50 percent power output. The controller will cycle on and off while within the proportional band.

- **Proportional band** — A temperature band expressed in degrees (if the input is temperature), or counts (if the input is process) from the set point in which the controllers' proportioning action takes place. The wider the proportional band the greater the area around the setpoint in which the proportional action takes place. It is sometimes referred to as gain, which is the reciprocal of proportional band.

- Integral, also known as reset, is a function which adjusts the proportional bandwidth with respect to the setpoint, to compensate for offset (droop) from setpoint, that is, it adjusts the controlled temperature to setpoint after the system stabilizes.

- Derivative, also known as rate, senses the rate of rise or fall of system temperature and automatically adjusts the proportional band to minimize overshoot or undershoot.

A PID (three-mode) controller is capable of exceptional control stability when properly tuned and used. The operator can achieve the fastest response time and smallest overshoot by following these instructions carefully. The information for tuning this three mode controller may be different from other controller tuning procedures. Normally an AUTO PID tuning feature will eliminate the necessity to use this manual tuning procedure for the primary output, however, adjustments to the AUTO PID values may be made if desired.

After the controller is installed and wired:

1. Apply power to the controller.
2. Disable the control outputs. (Push enter twice)
3. Program the controller for the correct input type (See Quick Start Manual).
4. Enter desired value for setpoint 1
5. For time proportional relay output, set the cycle time to 10 seconds or greater.
   - Press MENU until OUT1 is displayed.
   - Press ENTER to access control output 1 submenu.
   - Press MENU until cycle time is displayed.
   - Press ENTER to access cycle time setting.
   - Use MAX and MIN to set new cycle time value.
   - Press ENTER when finished.

6. Set prop band in degrees to 5% of setpoint 1. (If setpoint 1 = 100, enter 0005. Prop band = 95 to 110). Note: Micro-Infinity takes degrees (if input is temperature) / counts (if input is process) as Proportional Band value.

- If ID is disabled: - Press MENU 1 time from run mode to get to setpoint 1; confirm SP1 LED is flashing. - Use MAX and MIN to set new setpoint value.
- If ID is enabled: - Press MENU until Set Point is displayed. - Press ENTER to access setpoint 1 setting. - Use MAX and MIN to set new setpoint value.
- Press ENTER to stored setting when finished.

7. Set reset and rate to 0.

- Press MENU until OUT1 is displayed.
- Press ENTER to access control output 1 submenu.
- Press MENU until autopid is displayed.
- Press ENTER to access autopid setting.
- Press MAX to disable autopid; press ENTER when done.
- Press MENU until Reset Setup is displayed.
- Press ENTER to access Reset setting.
- Use MAX and MIN to set Reset to 0; press ENTER to store the new setting.
- Display advances to Rate Setup.
- Press ENTER to access Rate setting.
- Use MAX and MIN to set Rate to 0; press ENTER to store the new setting.
- Press MIN 2 times to return to run-mode. Should the unit reset, press ENTER twice to put it into stand-by mode.

NOTE: On units with dual three-mode outputs, the primary and secondary proportional parameter is independently set and may be tuned separately. The procedure used in this section is for a HEATING primary output. A similar procedure may be used for a primary COOLING output or a secondary COOLING output.

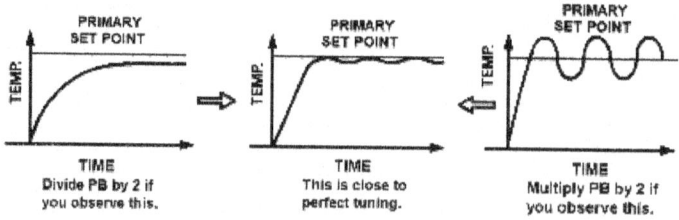

Fig. : Temperature Oscillations.

A. TUNING OUTPUTS FOR HEATING CONTROL

1. Enable the OUTPUT (Press Enter) and start the process.

2. The process should be run at a setpoint that will allow the temperature to stabilize with heat input required.

3. With RATE and RESET turned OFF, the temperature will stabilize with a steady state deviation, or droop, between the setpoint and the actual temperature. Carefully note whether or not there are regular cycles or oscillations in this temperature by observing the measurement on the display. (An oscillation may be as long as 30 minutes). 3. The tuning procedure is easier to follow if you use a recorder to monitor the process temperature.

4. If there are no regular oscillations in the temperature, divide the PB by 2. Allow the process to stabilize and check for temperature oscillations. If there are still no oscillations, divide the PB by 2 again. Repeat until cycles or oscillations are obtained. Proceed to Step 5.

5. If oscillations are observed immediately, multiply the PB by 2. Observe the resulting temperature for several minutes. If the oscillations continue, increase the PB by factors of 2 until the oscillations stop.

6. The PB is now very near its critical setting. Carefully increase or decrease the PB setting until cycles or oscillations just appear in the temperature recording.

7. If no oscillations occur in the process temperature even at the minimum PB setting skip Steps 6 through 15 below and proceed to paragraph B.

8. Read the steady-state deviation, or droop, between setpoint and actual temperature with the "critical" PB setting you have achieved. (Because the temperature is cycling a bit, use the average temperature.)

9. Measure the oscillation time, in minutes, between neighboring peaks or valleys. This is most easily accomplished with a chart recorder, but a measurement can be read at one minute intervals to obtain the timing.

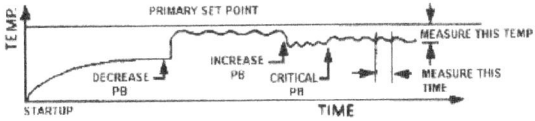

Fig. : Oscillation Time.

1. Now, increase the PB setting until the temperature deviation, or droop, increases 65%. The desired final temperature deviation can be calculated by multiplying the initial temperature deviation achieved with the CRITICAL PB setting by 1.65. Try several trial-and-error settings of the PB control until the desired final temperature deviation is achieved.

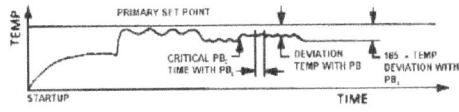

Fig. : Calculating Final Temperature Deviation.

2. You have now completed all the necessary measurements to obtain optimum performance from the Controller. Only two more adjustments are required — RATE and RESET.

3. Using the oscillation time measured in Step 7, calculate the value for RESET in repeats per minutes as follows:

$$RESET = (5/8) \times To$$

Where To = Oscillation Time in Seconds. Enter the value for RESET in OUT 1 (follow the same procedure as outlined in preparation section, step 7 to set RESET).

4. Again using the oscillation time measured in Step 7, calculate the value for RATE in minutes as follows:

$$RATE = To\ 10$$

Where T = Oscillation Time in Seconds. Enter this value for RATE in OUT 1 (follow the same procedure as outline in preparation section, step 7 to set RATE).

5. If overshoot occurred, it can be reduced by increasing the proportional band and the RESET time. When changes are made in the RESET value, a corresponding change should also be made in the RATE adjustment so that the RATE value is equal to:

$$RATE = (4/25) \times RESET$$

6. Several setpoint changes and consequent Prop Band, RESET and RATE time adjustments may be required to obtain the proper balance between "RESPONSE TIME" to a system upset and "SETTLING TIME". In general, fast response is accompanied by larger overshoot and consequently shorter time for the process to "SETTLE OUT". Conversely, if the response is slower, the process tends to slide into the final value with little or no overshoot. The requirements of the system dictate which action is desired.

7. When satisfactory tuning has been achieved, the cycle time should be increased to save contactor life (applies to units with time proportioning outputs only. Increase the cycle time as much as possible without causing oscillations in the measurement due to load cycling.

Tuning Procedure when no Oscillations are Observed

1. Measure the steady-state deviation, or droop, between setpoint and actual temperature with minimum PB setting.

2. Increase the PB setting until the temperature deviation (droop) increases 65%.

3. Set the RESET in OUT1 to a low value (50 secs). Set the RATE to zero (0 secs). At this point, the measurement should stabilize at the setpoint temperature due to reset action.

4.   Since we were not able to determine a critical oscillation time, the optimum settings of the reset and rate adjustments must be determined by trial and error. After the temperature has stabilized at setpoint, increase the setpoint temperature setting by 10 degrees. Observe the overshoot associated with the rise in actual temperature. Then return the setpoint setting to its original value and again observe the overshoot associated with the actual temperature change.

5.   Excessive overshoot implies that the Prop Band and/or RESET are set too low, and/or RATE value is set too high. Overdamped response (no overshoot) implies that the Prop Band and/or RESET is set too high, and/or RATE value is set too low. Where improved performance is required, change one tuning parameter at a time and observe its effect on performance when the setpoint is changed. Make incremental changes in the parameters until the performance is optimized.

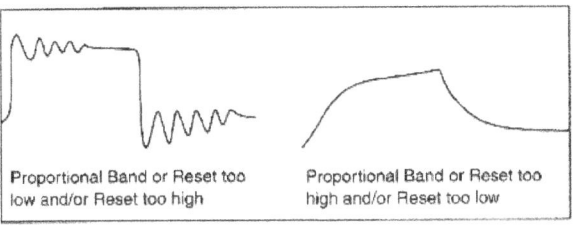

Fig. : Setting Reset and/or Rate PV.

1.   When satisfactory tuning has been achieved, the cycle time should be increased to save contactor life (applies to units with time proportioning outputs only.). Increase the cycle time as much as possible without causing oscillations in the measurement due to load cycling.

Tuning the Primary Output for Cooling Control

The same procedure is used as defined for heating. The process should be run at a setpoint that requires cooling control before the temperature will stabilize.

Simplified Tuning Procedure for PID Controllers

The following procedure is a graphical technique of analyzing a process response curve to a step input. It is much easier with a strip chart recorder reading the process variable (PV).

1.   Starting from a cold start (PV at ambient), apply full power to the process without the controller in the loop, *i.e.*, open loop. Record this starting time.

2.   After some delay (for heat to reach the sensor), the PV will start to rise. After more of a delay, the PV will reach a maximum rate of change (slope). Record the time that this maximum slope occurs, and the PV at which it occurs. Record the maximum slope in degrees per minute. Turn off system power.

3. Draw a line from the point of maximum slope back to the ambient temperature axis to obtain the lumped system time delay Td . The time delay may also be obtained by the equation: Td = time to max. slope – (PV at max. slope – Ambient)/max. slope

4. Apply the following equations to yield the PID parameters: Pr. Band = Td x max. slope Reset = Td/0.4 secs. Rate = 0.4 x Td minutes

5. Restart the system and bring the process to setpoint with the controller in the loop and observe response. If the response has too much overshoot, or is oscillating, then the PID parameters can be changed (slightly, one at a time, and observing process response) in the following directions: 5

Example: The chart recording in Figure 5 was obtained by applying full power to an oven. The chart scales are 10°F/cm, and 5 min/cm. The controller range is -200 - 900°F, or a span of 1100°F. Maximum slope = 18°F/5 minutes = 3.6°F/minutes. Time delay = Td = approximately 7 minutes.

Proportional Band = 7 minutes x 3.6°F / minutes = 25.2°F.

Note: Prop Band in Micro-Infinity is set in degrees/ counts. Reset = 7/.04 minutes = 17.5 min. or 1050 secs. Note: Reset in Micro-Infinity is specified in seconds Rate = 0.4 x 7 minutes = 2.8 min. or 168 secs.

Set Prop Band to: 025.0; Set Reset to: 1050 Set Rate to: 168 Follow step 6 and 7 of the preparation section to set new values for Prop Band, Reset, and Rate.

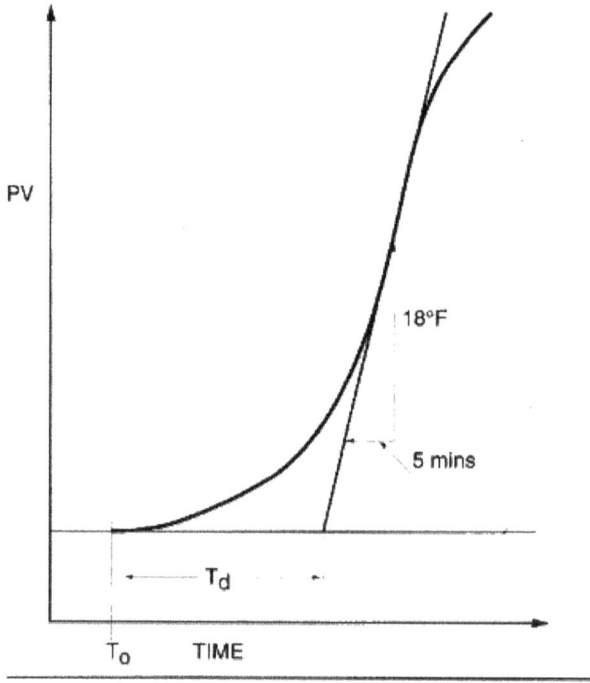

Figure.

PID TUNING CLASSICAL

Currently, more than half of the controllers used in industry are PID controllers. In the past, many of these controllers were analog; however, many of today's controllers use digital signals and computers. When a mathematical model of a system is available, the parameters of the controller can be explicitly determined. However, when a mathematical model is unavailable, the parameters must be determined experimentally. Controller tuning is the process of determining the controller parameters which produce the desired output. Controller tuning allows for optimization of a process and minimizes the error between the variable of the process and its set point.

Types of controller tuning methods include the trial and error method, and process reaction curve methods. The most common classical controller tuning methods are the Ziegler-Nichols and Cohen-Coon methods. These methods are often used when the mathematical model of the system is not available. The Ziegler-Nichols method can be used for both closed and open loop systems, while Cohen-Coon is typically used for open loop systems. A closed-loop control system is a system which uses feedback control. In an open-loop system, the output is not compared to the input.

The equation below shows the PID algorithm as discussed in the previous PID Control section.

$$u(t) = K_c \left(\in(t) + \frac{1}{\tau_i} \int_0^t \in(t')\,dt' + \tau_d \frac{d\in(t)}{dt} \right) + b$$

u is the control signal

ε is the difference between the current value and the set point.

K_c is the gain for a proportional controller.

τ_i is the parameter that scales the integral controller.

τ_d is the parameter that scales the derivative controller.

t is the time taken for error measurement.

b is the set point value of the signal, also known as bias or offset.

The experimentally obtained controller gain which gives stable and consistent oscillations for closed loop systems, or the ultimate gain, is defined as K_u. K_c is the controller gain which has been corrected by the Ziegler-Nichols or Cohen-Coon methods, and can be input into the above equation. K_u is found experimentally by starting from a small value of K_c and adjusting upwards until consistent oscillations are obtained, as shown below. If the gain is too low, the output signal will be damped and attain equilibrium eventually after the disturbance occurs as shown below.

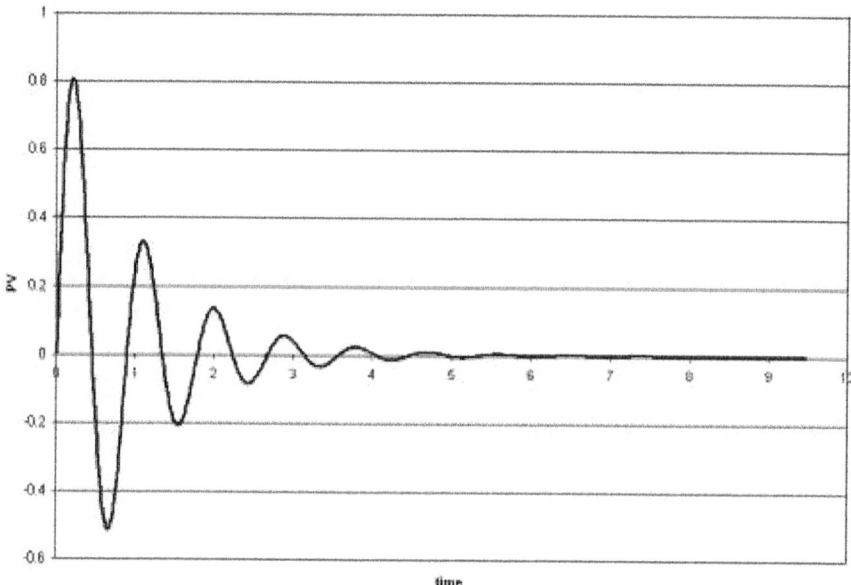

On the other hand, if the gain is too high, the oscillations become unstable and grow larger and larger with time as shown below.

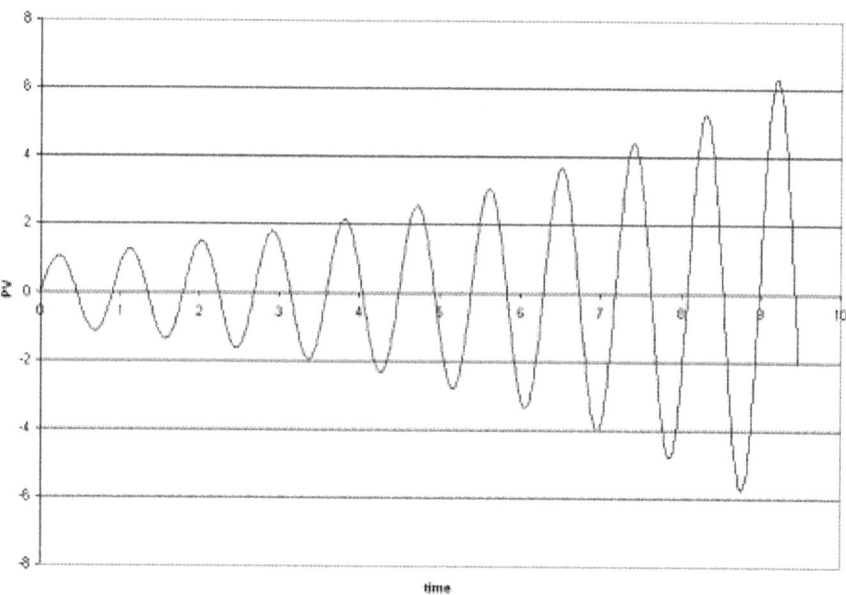

The process reaction curve method section shows the parameters required for open loop system calculations. The Ziegler-Nichols Method section shows how to find K_c, T_i, and T_d for open and closed loop systems, and the Cohen-Coon section shows an alternative way to find K_c, T_i, and T_d.

Open loop systems typically use the quarter decay ratio (QDR) for oscillation dampening. This means that the ratio of the amplitudes of the first overshoot to the second overshoot is 4:1.

Trial and Error

The trial and error tuning method is based on guess-and-check. In this method, the proportional action is the main control, while the integral and derivative actions refine it. The controller gain, K_c, is adjusted with the integral and derivative actions held at a minimum, until a desired output is obtained.

Below are some common values of K_c, T_i, and T_d used in controlling flow, levels, pressure or temperature for trial and error calculations.

Flow

P or PI control can be used with low controller gain. Use PI control for more accuracy with high integration activity. Derivative control is not considered due to the rapid fluctuations in flow dynamics with lots of noise.

$K_c = 0.4\text{-}0.65$

$T_i = 6s$

Level

P or PI control can be used, although PI control is more common due to inaccuracies incurred due to offsets in P-only control. Derivative control is not considered due to the rapid fluctuations in flow dynamics with lots of noise.

The following P only setting is such that the control valve is fully open when the vessel is 75% full and fully closed when 25% full, being half open when 50% filled.

$K_c = 2$

Bias b = 50%

Set point = 50%

For PI control:

$K_c = 2\text{-}20$

$T_i = 1\text{-}5$ min

Pressure

Tuning here has a large range of possible values of K_c and T_i for use in PI control, depending on if the pressure measurement is in liquid or gas phase.

Liquid

$K_c = 0.5\text{-}2$

T_i = 6-15 s

Gas

K_c = 2-10

T_i = 2-10 min

Temperature

Due to the relatively slow response of temperature sensors to dynamic temperature changes, PID controllers are used.

K_c = 2-10

T_i = 2-10 min

T_d = 0-5 min

Process Reaction Curve

In this method, the variables being measured are those of a system that is already in place. A disturbance is introduced into the system and data can then be obtained from this curve. First the system is allowed to reach steady state, and then a disturbance, X_o, is introduced to it. The percentage of disturbance to the system can be introduced by a change in either the set point or process variable. For example, if you have a thermometer in which you can only turn it up or down by 10 degrees, then raising the temperature by 1 degree would be a 10% disturbance to the system.

These types of curves are obtained in open loop systems when there is no control of the system, allowing the disturbance to be recorded. The process reaction curve method usually produces a response to a step function change for which several parameters may be measured which include: transportation lag or dead time, τ_{dead}, the time for the response to change, τ, and the ultimate value that the response reaches at steady-state, M_u.

τ_{dead} = transportation lag or dead time: the time taken from the moment the disturbance was introduced to the first sign of change in the output signal

τ = the time for the response to occur

X_o = the size of the step change

M_u = the value that the response goes to as the system returns to steady-state

$$R = \frac{\tau_{dead}}{\tau}$$

$$K_o = \frac{X_o}{M_u} \frac{\tau}{\tau_{deat}}$$

An example for determining these parameters for a typical process response curve to a step change is shown below.

In order to find the values for τ_{dead} and τ, a line is drawn at the point of inflection that is tangent to the response curve and then these values are found from the graph.

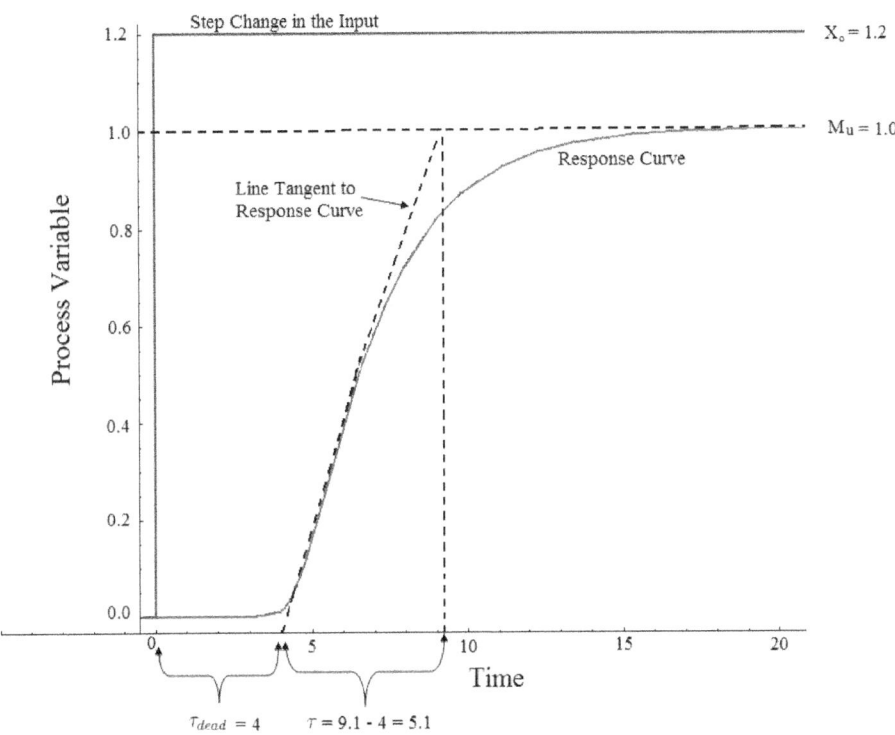

Ziegler-Nichols Method

In the 1940's, Ziegler and Nichols devised two empirical methods for obtaining controller parameters. Their methods were used for non-first order plus dead time situations, and involved intense manual calculations. With improved optimization software, most manual methods such as these are no longer used. However, even with computer aids, the following two methods are still employed today, and are considered among the most common:

Ziegler-Nichols Closed-loop Tuning Method

The Ziegler-Nichols closed-loop tuning method allows you to use the ultimate gain value, K_u, and the ultimate period of oscillation, P_u, to calculate K_c. It is a simple method of tuning PID controllers and can be refined to give better approximations of the controller. You can obtain the controller constants K_c, T_i, and T_d in a system with feedback. The Ziegler-Nichols closed-loop tuning method is limited to tuning processes that cannot run in an open-loop environment.

Determining the ultimate gain value, K_u, is accomplished by finding the value of the proportional-only gain that causes the control loop to oscillate indefinitely at steady state. This means that the gains from the I and D controller are set to zero so that the influence of P can be determined. It tests the robustness of the K_c value so that it is optimized for the controller. Another important value associated with this proportional-only control tuning method is the ultimate period (P_u). The ultimate period is the time required to complete one full oscillation while the system is at steady state. These two parameters, K_u and P_u, are used to find the loop-tuning constants of the controller (P, PI, or PID). To find the values of these parameters, and to calculate the tuning constants, use the following procedure:

Closed Loop (Feedback Loop)

1. Remove integral and derivative action. Set integral time (T_i) to 999 or its largest value and set the derivative controller (T_d) to zero.
2. Create a small disturbance in the loop by changing the set point. Adjust the proportional, increasing and/or decreasing, the gain until the oscillations have constant amplitude.
3. Record the gain value (K_u) and period of oscillation (P_u).

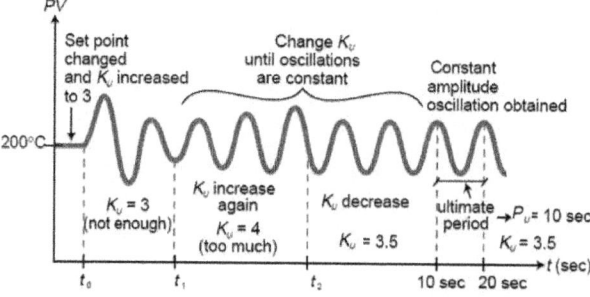

Fig. : System tuned using the Ziegler-Nichols closed-loop tuning method

4. Plug these values into the Ziegler-Nichols closed loop equations and determine the necessary settings for the controller.

Table : Closed-Loop Calculations of K_c, T_i, T_d

	K_c	T_I	T_D
P	$K_u/2$		
PI	$K_u/2.2$	$P_u/1.2$	
PID	$K_u/1.7$	$P_u/2$	$P_u/8$

Advantages

1. Easy experiment; only need to change the P controller
2. Includes dynamics of whole process, which gives a more accurate picture of how the system is behaving

Disadvantages

1. Experiment can be time consuming
2. Can venture into unstable regions while testing the P controller, which could cause the system to become out of control

Ziegler-Nichols Open-Loop Tuning Method or Process Reaction Method:

This method remains a popular technique for tuning controllers that use proportional, integral, and derivative actions. The Ziegler-Nichols open-loop method is also referred to as a process reaction method, because it tests the open-loop reaction of the process to a change in the control variable output. This basic test requires that the response of the system be recorded, preferably by a plotter or computer. Once certain process response values are found, they can be plugged into the Ziegler-Nichols equation with specific multiplier constants for the gains of a controller with either P, PI, or PID actions.

Open Loop (Feed Forward Loop)

To use the Ziegler-Nichols open-loop tuning method, you must perform the following steps:

1. Make an open loop step test
2. From the process reaction curve determine the transportation lag or dead time, τ_{dead}, the time constant or time for the response to change, τ, and the ultimate value that the response reaches at steady-state, M_u, for a step change of Xo.

$$K_o = \frac{X_o}{M_u} \frac{\tau}{\tau_{deat}}$$

3. Determine the loop tuning constants. Plug in the reaction rate and lag time values to the Ziegler-Nichols open-loop tuning equations for the appropriate controller — P, PI, or PID — to calculate the controller constants. Use the table below.

Table: Open-Loop Calculations of K_c, T_i, T_d

	K_c	T_i	T_d
P	K_0		
PI	$0.9K_0$	$3.3\tau_{dead}$	
PID	$1.2K_0$	$2\tau_{dead}$	$0.5\tau_{dead}$

Advantages

1. Quick and easier to use than other methods
2. It is a robust and popular method

3. Of these two techniques, the Process Reaction Method is the easiest and least disruptive to implement

Disadvantages

1. It depends upon purely proportional measurement to estimate I and D controllers.

2. Approximations for the K_c, T_i, and T_d values might not be entirely accurate for different systems.

3. It does not hold for I, D and PD controllers

Cohen-Coon Method

The Cohen-Coon method of controller tuning corrects the slow, steady-state response given by the Ziegler-Nichols method when there is a large dead time (process delay) relative to the open loop time constant; a large process delay is necessary to make this method practical because otherwise unreasonably large controller gains will be predicted. This method is only used for first-order models with time delay, due to the fact that the controller does not instantaneously respond to the disturbance (the step disturbance is progressive instead of instantaneous).

The Cohen-Coon method is classified as an 'offline' method for tuning, meaning that a step change can be introduced to the input once it is at steady-state. Then the output can be measured based on the time constant and the time delay and this response can be used to evaluate the initial control parameters.

For the Cohen-Coon method, there are a set of pre-determined settings to get minimum offset and standard decay ratio of 1/4(QDR). A 1/4(QDR) decay ratio refers to a response that has decreasing oscillations in such a manner that the second oscillation will have 1/4 the amplitude of the first oscillation. These settings are shown in Table.

Table: Standard recommended equations to optimize Cohen Coon predictions.

	Kc	Ti	Td
P	(P/NL)*(1+(R/3))		
PI	(P/NL)*(0.9+(R/12))	L*(30+3R) / (9+20R)	
PID	(P/NL)*(1.33+(R/4))	L*(30+3R) / (9+20R)	4L / (11+2R)

where the variables P, N, and L are defined below.

P	Percent Change of Input
N	Percent Change of Output/τ
L	τ_{dead}
R	τ_{dead}/τ

Alternatively, K_0 can be used instead of (P/NL). K_0, τ, and τ_{dead} are defined in process reaction curve section. An example using these parameters is shown here.

The process in Cohen-Coon turning method is the following:

1. Wait until the process reaches steady state.

2. Introduce a step change in the input.

3. Based on the output, obtain an approximate first order process with a time constant τ delayed by τ_{dead} units from when the input step was introduced.

The values of τ and τ_{dead} can be obtained by first recording the following time instances:

t0 = time at input step start point t2 = time when reaches half point t3 = time when reaches 63.2% point

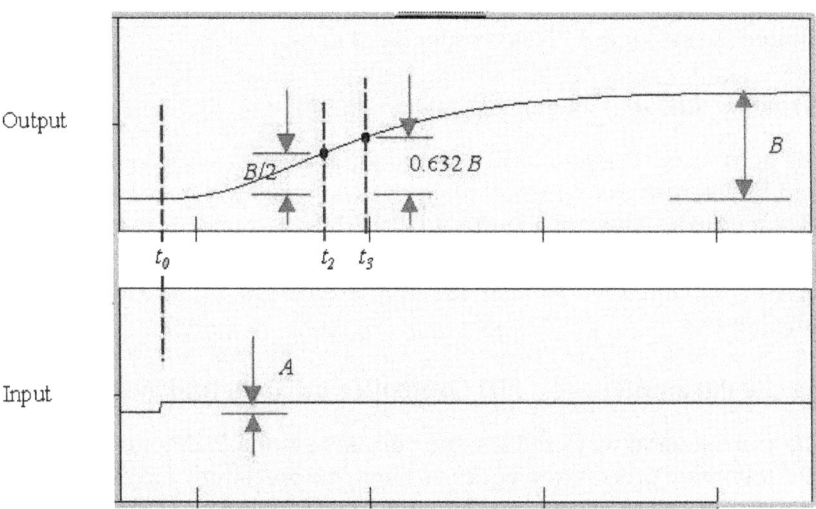

4. Using the measurements at t_0, t_2, t_3, A and B, evaluate the process parameters τ, τ_{dead}, and K_o.

5. Find the controller parameters based on τ, τ_{dead}, and K_o.

Advantages

1. Used for systems with time delay.

2. Quicker closed loop response time.

Disadvantages and Limitations

1. Unstable closed loop systems.

2. Can only be used for first order models including large process delays.

3. Offline method.

4. Approximations for the K_c, τ_i, and τ_d values might not be entirely accurate for different systems.

Other Methods

These are other common methods that are used, but they can be complicated and aren't considered classical methods, so they are only briefly discussed.

Internal Model Control

The Internal Model Control (IMC) method was developed with robustness in mind. The Ziegler-Nichols open loop and Cohen-Coon methods give large controller gain and short integral time, which isn't conducive to chemical engineering applications. The IMC method relates to closed-loop control and doesn't have overshooting or oscillatory behavior. The IMC methods however are very complicated for systems with first order dead time.

Auto Tune Variation

The auto-tune variation (ATV) technique is also a closed loop method and it is used to determine two important system constants (P_u and K_u for example). These values can be determined without disturbing the system and tuning values for PID are obtained from these. The ATV method will only work on systems that have significant dead time or the ultimate period, P_u, will be equal to the sampling period.

Tuning the Parameters of a PID Controller on an Actual System

There are several ways to tune the parameters of a PID controller. They involve the following procedures. For each, name the procedure and explain how the given measured information is used to pick the parameters of the PID controller.

PID THEORY EXPLAINED

Proportional-Integral-Derivative (PID) control is the most common control algorithm used in industry and has been universally accepted in industrial control. The popularity of PID controllers can be attributed partly to their robust performance in a wide range of operating conditions and partly to their functional simplicity, which allows engineers to operate them in a simple, straightforward manner.

As the name suggests, PID algorithm consists of three basic coefficients; proportional, integral and derivative which are varied to get optimal response. Closed loop systems, the theory of classical PID and the effects of tuning a closed loop control system are discussed in this paper. The PID toolset in LabVIEW and the ease of use of these VIs is also discussed.

Control System

The basic idea behind a PID controller is to read a sensor, then compute the desired actuator output by calculating proportional, integral, and derivative responses and summing those three components to compute the output. Before we start to define the parameters of a PID controller, we shall see what a closed loop system is and some of the terminologies associated with it.

Closed Loop System

In a typical control system, the process variable is the system parameter that needs to be controlled, such as temperature (°C), pressure (psi), or flow rate (liters/minute). A sensor is used to measure the process variable and provide feedback to the control system. The set point is the desired or command value for the process variable, such as 100 degrees Celsius in the case of a temperature control system. At any given moment, the difference between the process variable and the set point is used by the control system algorithm (compensator), to determine the desired actuator output to drive the system (plant).

For instance, if the measured temperature process variable is 100 °C and the desired temperature set point is 120 °C, then the actuator output specified by the control algorithm might be to drive a heater. Driving an actuator to turn on a heater causes the system to become warmer, and results in an increase in the temperature process variable. This is called a closed loop control system, because the process of reading sensors to provide constant feedback and calculating the desired actuator output is repeated continuously and at a fixed loop rate as illustrated in figure.

In many cases, the actuator output is not the only signal that has an effect on the system. For instance, in a temperature chamber there might be a source of cool air that sometimes blows into the chamber and disturbs the temperature. Such a term is referred to as disturbance. We usually try to design the control system to minimize the effect of disturbances on the process variable.

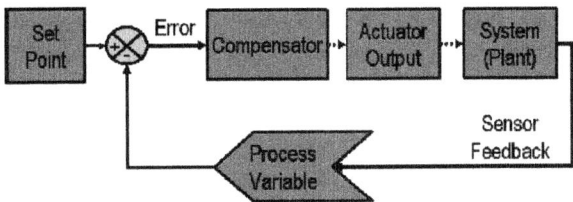

Fig. : Block diagram of a typical closed loop system.

Defintion of Terminlogies

The control design process begins by defining the performance requirements. Control system performance is often measured by applying a step function as the set point command variable, and then measuring the response of the process variable. Commonly, the response is quantified by measuring defined waveform characteristics. Rise Time is the amount of time the system takes to go from 10% to 90% of the steady-state, or final, value. Percent Overshoot is the amount that the process variable overshoots the final value, expressed as a percentage of the final value. Settling time is the time required for the process variable to settle to within a certain percentage (commonly 5%) of the final value. Steady-State Error is the final difference between the process variable and set point. Note that the exact definition of these quantities will vary in industry and academia.

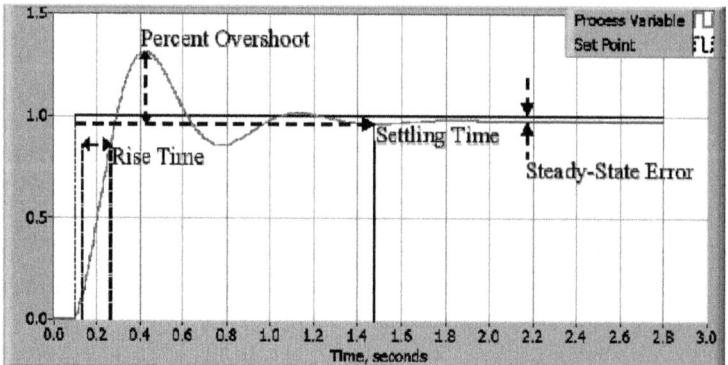

Fig. : Response of a typical PID closed loop system.

After using one or all of these quantities to define the performance requirements for a control system, it is useful to define the worst case conditions in which the control system will be expected to meet these design requirements. Often times, there is a disturbance in the system that affects the process variable or the measurement of the process variable. It is important to design a control system that performs satisfactorily during worst case conditions. The measure of how well the control system is able to overcome the effects of disturbances is referred to as the disturbance rejection of the control system.

In some cases, the response of the system to a given control output may change over time or in relation to some variable. A nonlinear system is a system in which the control parameters that produce a desired response at one operating point might not produce a satisfactory response at another operating point. For instance, a chamber partially filled with fluid will exhibit a much faster response to heater output when nearly empty than it will when nearly full of fluid. The measure of how well the control system will tolerate disturbances and nonlinearities is referred to as the robustness of the control system.

Some systems exhibit an undesirable behavior called deadtime. Deadtime is a delay between when a process variable changes, and when that change can be observed. For instance, if a temperature sensor is placed far away from a cold water fluid inlet valve, it will not measure a change in temperature immediately if the valve is opened or closed. Deadtime can also be caused by a system or output actuator that is slow to respond to the control command, for instance, a valve that is slow to open or close. A common source of deadtime in chemical plants is the delay caused by the flow of fluid through pipes.

Loop cycle is also an important parameter of a closed loop system. The interval of time between calls to a control algorithm is the loop cycle time. Systems that change quickly or have complex behavior require faster control loop rates.

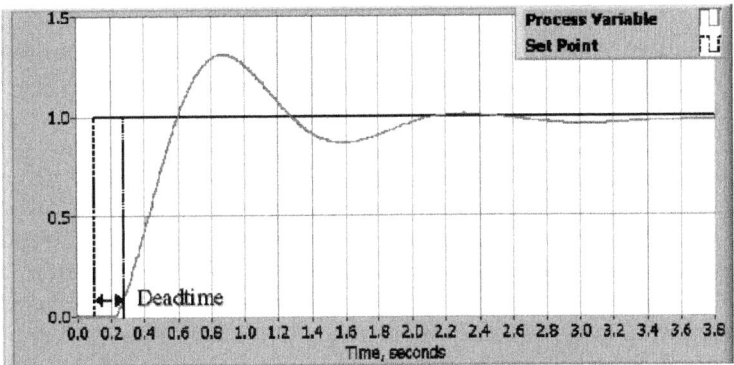

Fig. : Response of a closed loop system with deadtime.

Once the performance requirements have been specified, it is time to examine the system and select an appropriate control scheme. In the vast majority of applications, a PID control will provide the required results

PID Theory

Proportional Response

The proportional component depends only on the difference between the set point and the process variable. This difference is referred to as the Error term. The proportional gain (K_c) determines the ratio of output response to the error signal. For instance, if the error term has a magnitude of 10, a proportional gain of 5 would produce a proportional response of 50. In general, increasing the proportional gain will increase the speed of the control system response. However, if the proportional gain is too large, the process variable will begin to oscillate. If K_c is increased further, the oscillations will become larger and the system will become unstable and may even oscillate out of control.

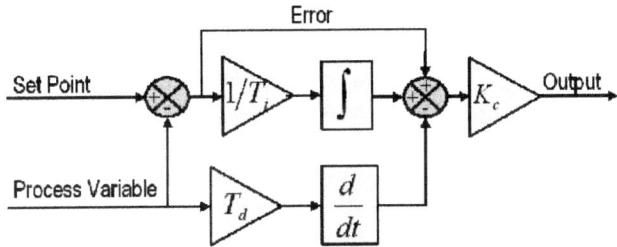

Fig. : Block diagram of a basic PID control algorithm.

Integral Response

The integral component sums the error term over time. The result is that even a small error term will cause the integral component to increase slowly. The integral response will continually increase over time unless the error is zero, so the effect is to drive the Steady-State error to zero. Steady-State error is the final difference between the process variable and set point. A phenomenon called integral windup results when integral action saturates a controller without the controller driving the error signal toward zero.

Derivative Response

The derivative component causes the output to decrease if the process variable is increasing rapidly. The derivative response is proportional to the rate of change of the process variable. Increasing the derivative time (T_d) parameter will cause the control system to react more strongly to changes in the error term and will increase the speed of the overall control system response. Most practical control systems use very small derivative time (T_d), because the Derivative Response is highly sensitive to noise in the process variable signal. If the sensor feedback signal is noisy or if the control loop rate is too slow, the derivative response can make the control system unstable

Tuning

The process of setting the optimal gains for P, I and D to get an ideal response from a control system is called tuning. There are different methods of tuning of which the "guess and check" method and the Ziegler Nichols method will be discussed.

The gains of a PID controller can be obtained by trial and error method. Once an engineer understands the significance of each gain parameter, this method becomes relatively easy. In this method, the I and D terms are set to zero first and the proportional gain is increased until the output of the loop oscillates. As one increases the proportional gain, the system becomes faster, but care must be taken not make the system unstable. Once P has been set to obtain a desired fast response, the integral term is increased to stop the oscillations.

The integral term reduces the steady state error, but increases overshoot. Some amount of overshoot is always necessary for a fast system so that it could respond to changes immediately. The integral term is tweaked to achieve a minimal steady state error. Once the P and I have been set to get the desired fast control system with minimal steady state error, the derivative term is increased until the loop is acceptably quick to its set point. Increasing derivative term decreases overshoot and yields higher gain with stability but would cause the system to be highly sensitive to noise. Often times, engineers need to tradeoff one characteristic of a control system for another to better meet their requirements.

The Ziegler-Nichols method is another popular method of tuning a PID controller. It is very similar to the trial and error method wherein I and D are set to zero and P is increased until the loop starts to oscillate. Once oscillation starts, the critical gain K_c and the period of oscillations P_c are noted. The P, I and D are then adjusted as per the tabular column shown below.

Table: Ziegler-Nichols tuning, using the oscillation method.

Control	P	Ti	Td
P	0.5Kc	-	-
PI	0.45Kc	Pc/1.2	-
PID	0.60Kc	0.5Pc	Pc/8

NI LabVIEW and PID

LabVIEW PID toolset features a wide array of VIs that greatly help in the design of a PID based control system. Control output range limiting, integrator anti-windup and bumpless controller output for PID gain changes are some of the salient features of the PID VI. The PID Advanced VI includes all the features of the PID VI along with non-linear integral action, two degree of freedom control and error-squared control.

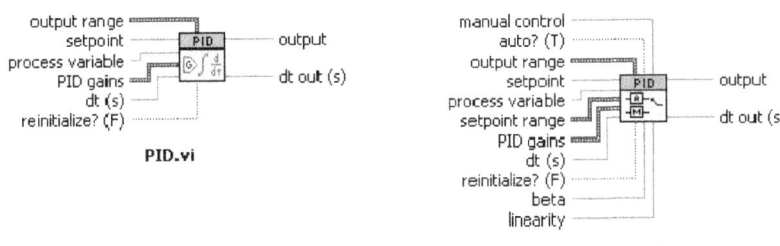

Fig. : VIs from the PID controls palette of LabVIEW.

PID palette also features some advanced VIs like the PID Autotuning VI and the PID Gain Schedule VI. The PID Autotuning VI helps in refining the PID parameters of a control system. Once an educated guess about the values of P, I and D have been made, the PID Autotuning VI helps in refining the PID parameters to obtain better response from the control system.

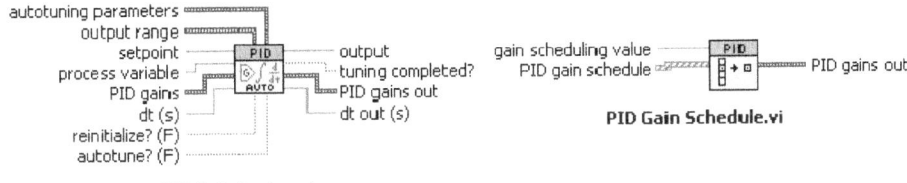

Fig. : Advanced VIs from the PID controls palette of LabVIEW.

The reliability of the controls system is greatly improved by using the LabVIEW Real Time module running on a real time target. National Instruments provides the new M Series Data Acquisition boards which provide higher accuracy and better performance than an average control system.

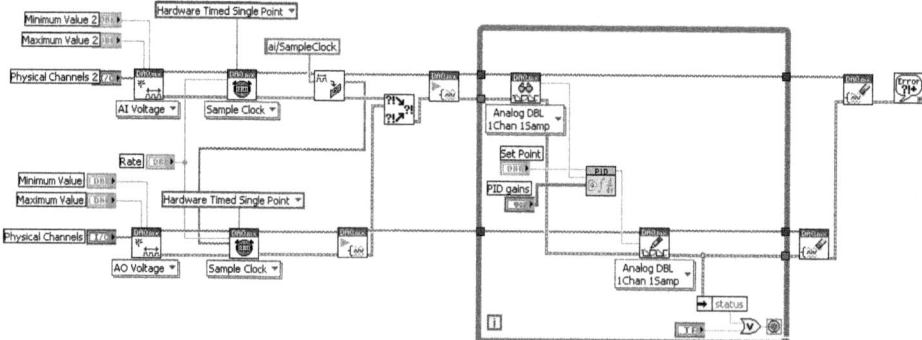

Fig. : A typical LabVIEW VI showing PID control with a plug-in
NI data acquisition device.

The tight integration of these M Series boards with LabVIEW minimizes the development time involved and greatly increases the productivity of any engineer. Figure shows a typical VI in LabVIEW showing PID control using NI-DAQmx API of M series devices.

THE BASICS OF TUNING PID LOOPS

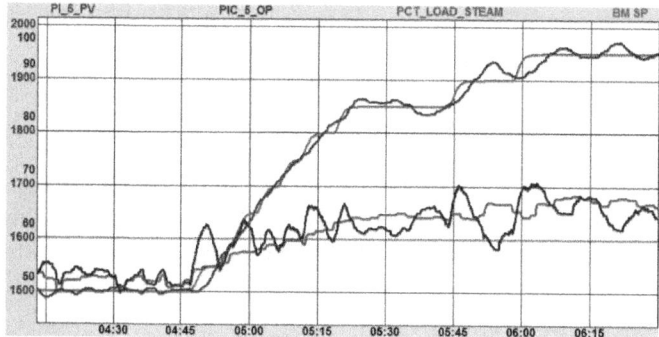

The art of **tuning a PID loop** is to have it adjust its OP to move the PV as quickly as possible to the SP (**responsive**), minimize **overshoot**, and then hold the PV steady at the SP without excessive OP changes (**stable**).

First, Some Definitions

- **PID** = Proportional, Integral, Derivative algorithm. This is not a P&ID, which is a Piping (or Process) and Instrumentation Diagram.
- **PV** = Process Variable - a quantity used as a feedback, typically measured by an instrument. Also sometimes called "MV" - Measured Value.
- **SP** = SetPoint - the desired value for the PV.
- **OP** = OutPut - a signal to a device that can change the PV - frequently a valve, damper, or a pump speed reference. Also sometimes called "CV" - Controlled Value.

- **Overshoot** = when the PV moves further past the SP than desired.
- A PID loop in **manual** (as opposed to **automatic**) only changes its OP upon operator request.
- A loop in **remote** has its SP automatically adjusted by external logic. In **local** the SP is only changed by the operator. Some systems combine auto and remote into "cascade" mode.
- A **direct acting** PID loop increases its OP in response to increasing PV, while a **reverse acting** loop decreases its OP. "Normal" loops are reverse acting. Loops controlling level or pressure via a valve on an output, or temperature via cooling are generally direct acting – "backwards" loops.
- **Error** = the difference between PV and SP.

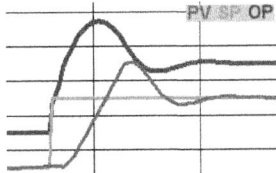

Fig. : Overshoot.

Basic Tuning Parameters of a PID Loop

Note: for demonstration purposes the charts below show the individual responses of the actions where the PV is **NOT** affected by the OP. Normally the PV would be affected by the change in OP & would therefore be brought back toward the SP as a result of the OP's response.

Gain

Also called **proportional band or P-gain**, the **gain** determines how much change the OP will make due to a change in error (from a PV change and / or an SP change). This mainly corrects the OP based on upsets as they happen. "Gain" implies that a larger number will have more effect. "Proportional band" implies the opposite. P-Gain = 100% / P-band.

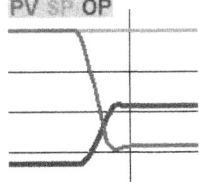

Fig. : Gain Only Response.

Reset

Also called **integral or I-gain**, the **reset** determines how much to change the OP over time due to the error (regardless of the direction of movement of the error). This brings a stable PV that is off SP toward the SP. Reset or I-gain implies that a larger number will have more effect. Integral implies the opposite. Reset [resets per minute] = 60 / Integral [seconds per reset].

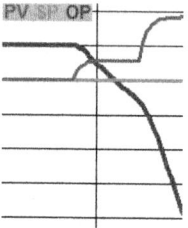

Fig. : Reset Only Response.

Preact

Also called **derivative or D-gain**, the **preact** determines how much to change the OP due from a change in direction of the error or PV. While acting on the PV, rather than the error, is an option in some loops, acting on PV is better because it is undesirable to bump the OP when the SP is changed. It is called preact because it allows the loop to "anticipate" upsets as they begin to happen and react quickly.

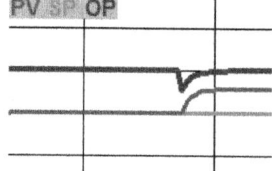

Fig. : Preact Only Response.

Actions Working Together

When the PV is approaching the SP, Proportional and Integral work in opposite directions to cause the PV of a properly tuned loop to get to the SP quickly without excessively overshooting. For a typical reverse-acting loop, the proportional will try to close the OP as the PV rises toward the SP, but the integral will try to open the OP because PV is below SP. As the PV gets closer to the SP, integral action decreases resulting in the PV smoothly decelerating into the SP.

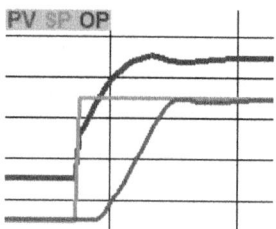

Fig. : Gain and Reset Together.

Initial Loop Tuning

The goal of tuning a PID loop is to make it stable, responsive and to minimize overshooting. These goals - especially the last two - conflict with each other. You must find a compromise between the goals which acceptably satisfies them all. Process requirements and physical limitations will determine the balance between amount of acceptable overshoot as well as the demand for responsiveness. The primary factors which dictate the limits to responsiveness of a loop are:

1. The amount of PV change resulting from an OP step change.

2. The time from when the OP changes until the change starts to be seen in the PV,

3. The time for the PV to reach its new level. Because the PV usually approaches its new level asymptotically, for that time constant we frequently use the time it takes the system's step response to reach $1-1/e$, or about 63% of the distance from the initial to its final value.

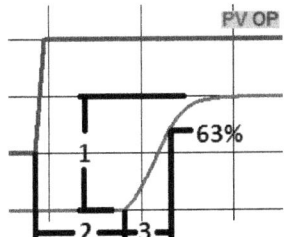

Fig. : Three Response Factors.

When beginning to tune a loop, first make sure you have a good trending package. Watch the PV, SP, OP and other outside variables you suspect might influence the PV together. Characterize the loop by making step changes in manual through the range of the OP and back, and make a note of these three factors. Stepping through the range in both directions is valuable to quantify the **linearity** and **hysteresis** of the system.

Linearity: If the same change in OP through the whole scale results in a similar change in PV at each point, the system is **linear**. But if a change on one part of the OP range results in more PV change than the same OP change in a different range, the system is **non-linear**.

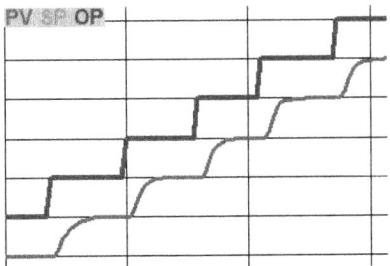

Fig. : Linear OP.

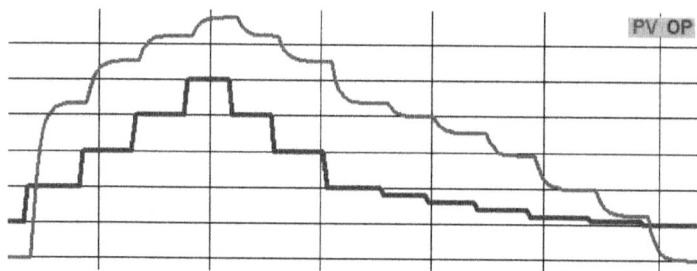

Fig. : Non Linear OP.

Hysteresis: Some devices will yield a different PV for the same OP depending on whether the OP went up or down to get there. A valve might allow 25 GPM through after moving from 20% to 30%, but 30 GPM after moving from 40% to 30%.

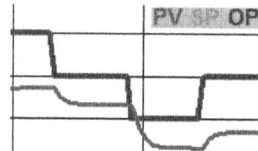

Fig. : OP Hysteresis.

Basic PV Categories: Particle and Bulk

Particle properties are those where a fluid in a pipe may have different properties in different areas, so that the fluid must be mixed or moved to change the property at the PV measurement point. Particle properties include temperature, pH, conductivity, *etc.* These tend to have a significant delay between a change in OP and the beginning of the change in PV (#2 above), and therefore might benefit from derivative action.

Bulk properties describe the state of the fluid as a whole so that it all changes everywhere in a pipe or vessel (for practical purposes) simultaneously. Examples: flow, level, and pressure. Generally these PVs begin to show the result of an OP change immediately (even if the time constant to complete the change is long) and do not need any derivative.

Another categorization of PVs: some (such as flow) increase when the OP increases and decrease when the OP decreases. These should be characterized as shown above - they typically need more integral and minimal P-gain. For others (such as level) the **direction** (rather than value) of the PV is relative to the OP. For the latter, a characterization is more subtle - you want to characterize the slope of the PV for various OPs instead of its value. These sometimes need moderate-to-high gain and less integral.

Starting Parameters

Loops where the PV changes quickly due to a change in OP (flow, or pressure or level in vessels with fast turnover) should have low P-gain (perhaps 0.2) and higher reset (1.5 – 10 rpm). Loops where the PV changes slowly, or changes its direction of movement due to change in OP (temperature and level in vessels with slow turnover) typically need high gain (3 – 100) and low reset (0.05 – 0.3).

These recommended starting parameters are based on the input and output ranges being the same. Some controllers handle tuning parameters based on percent of span, while others do not make this correction. If the spans are different, corrections would have to be made to the parameters themselves. For example, if the flow through a pipe can be from 0 – 10,000 gpm, and you are adjusting the speed of a VFD from 0-100%, the starting gain and reset would need to be 0.004 and 0.02 instead of 0.4 and 2.0.

There are numeric methods where the natural resonant frequency of a system is determined and parameters set accordingly, but **I've found an iterative, intuitive approach to be more useful**:

- Start with a low proportional and no integral or derivative.
- Double the proportional until it begins to oscillate, then halve it.
- Implement a small integral.
- Double the integral until it starts oscillating, then halve it.

That will get the constants close to where they need to be for fine adjustment. Don't hesitate to put the loop back in manual if the loop goes crazy or while studying the trend.

Fine Tuning

To achieve the goal of a responsive and stable loop with minimal overshoot, the tuning must be tested in response to upsets and at steady state. Upsets can be induced by:

- Changing the SP.
- Putting the loop in manual, changing the OP, then returning to automatic.
- Externally causing a change to the PV.

Once the PV has stabilized at its SP, upset it by stepping the SP. Even a very small change is useful here. The proportional will cause an immediate jump in OP, then the integral will cause the OP to continue ramping in the same direction. When the PV starts to move, the proportional will cause the OP to move back the other way, and the integral action will diminish as the PV approaches the SP. Overshoot is often caused by too much integral and/or not enough proportional.

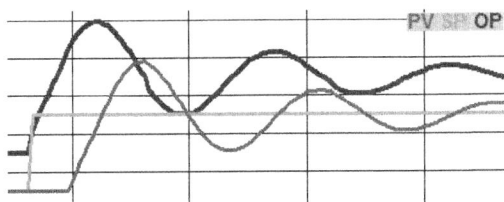

Fig. : Too Much Reset.

The OP needs to start moving back the other way well before the PV reaches the SP. The amount of time between the peak and the PV hitting the SP depends on the nature of the loop. If the peak comes too late, you need more proportional or less integral. If the peak comes too early, you need less proportional or more integral. An early peak will result in the PV leveling out before it reaches the SP, causing the OP and PV to swing on the way to their new steady-state values.

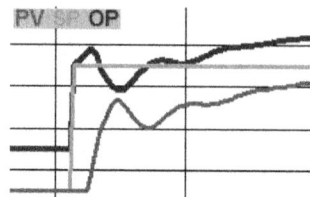

Fig. : Swinging to SP.

Once the loop is roughly tuned, put it in manual and change either the SP or OP, let it stabilize, then put it back in automatic. The loop will then move the PV to the SP, but with integral only - without the initial OP pulse from proportional. If the OP moves too slowly, you need more integral. A long delay between the change in OP and the end of the resulting change to PV dictates a lower integral value. Introduce derivative if you see that a bump in OP would be beneficial when the PV changes direction at the beginning of an upset.

This is an iterative method – every change in one parameter changes the ideal value for the other parameters. Go back and forth between upset methods and steady state stability, and make sure you check the tuning for the full range of possible SPs, If the system is non-linear, a loop that is stable at higher flows may swing wildly at lower flows, and a loop that is responsive at low flows may be sluggish at higher flows. A PID loop with a control deadband can sometimes achieve acceptable control despite this challenge.

Diagnosing the Cause of Oscillations

If the OP and PV peak at the same time, the oscillation is proportional-driven.

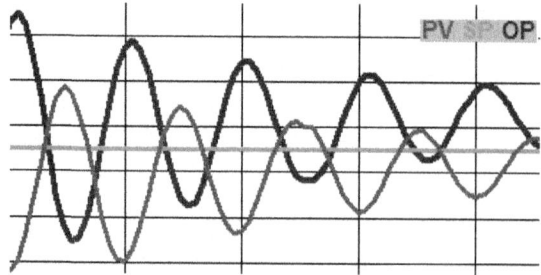

Fig. : Too Much Gain.

If the OP peaks when the PV is crossing its midpoint & visa versa (so that the PV and OP waves are 90° out of phase), the oscillation is integral-driven. Some oscillations are driven by other factors in the system - put the loop in manual to see if it continues to oscillate if you suspect the loop you are tuning is not causing the oscillation.

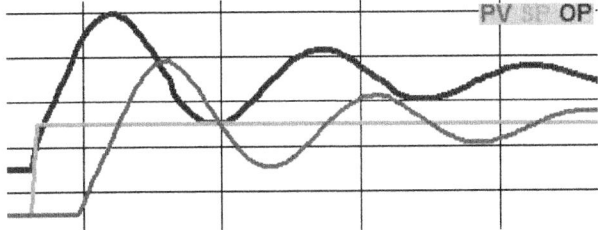

Fig. : Too Much Reset.

Sometimes oscillations are acceptable. For example, the goal of boiler drum level control is primarily to avoid tripping on either low or high level. A moderate amount of oscillation at steady state is a good trade-off to get enough additional responsiveness to avoid tripping following significant upsets.

Note that actuators (especially motorized valves) with deadband and/or limited duty cycle will ALWAYS swing when attached to a traditional PID loop regardless of tuning parameters. You can tell this is happening by looking at the trend – the PV will be flat while the OP is ramping down (due to integral), then the PV jumps to the other side of the SP, and the pattern reverses.

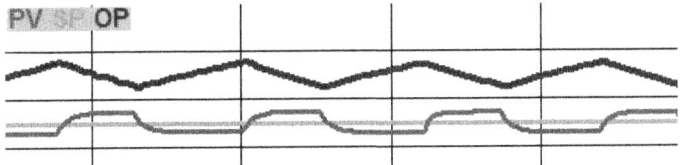

Fig. : Limited Duty Deadband Oscillation.

Balancing the Tuning Goals

A properly tuned loop balances the demands of stability, responsiveness and low overshoot. Tune the loop by adjusting the three tuning parameters so that the loop responds well in a variety of upset and steady-state situations.

LINEAR PARAMETER-VARYING CONTROL

Linear parameter-varying control (LPV control) deals with the control of linear parameter-varying systems, a class of nonlinear systems which can be modelled as parametrized linear systems whose parameters change with their state.

Gain Scheduling

In designing feedback controllers for dynamical systems a variety of modern, multivariable controllers are used. In general, these controllers are often designed at various operating points using linearized models of the system dynamics and are scheduled as a function of a parameter or parameters for operation at intermediate conditions. It is an approach for the control of non-linear systems that uses a family of linear controllers, each of which provides satisfactory control for a different operating point of the system. One or more observable variables, called the scheduling variables, are used to determine the current operating region of the system and to enable the appropriate linear controller. For example in case of aircraft control, a set of controllers are designed at different gridded locations of corresponding parameters such as AoA, Mach, dynamic pressure, CG *etc.* In brief, gain scheduling is a control design approach that constructs a nonlinear controller for a nonlinear plant by patching together a collection of linear controllers. These linear controllers are blended in real-time via switching or interpolation.

Scheduling multivariable controllers can be very tedious and time consuming task. A new paradigm is the linear parameter-varying (LPV) techniques which synthesize of automatically scheduled multivariable controller.

Drawbacks of Classical Gain Scheduling

- An important drawback of classical gain scheduling approach is that adequate performance and in some cases even stability is not guaranteed at operating conditions other than the design points.

- Scheduling multivariable controllers is often a tedious and time consuming task and it holds true especially in the field of aerospace control where the parameter dependency of controllers are large due to increased operating envelopes with more demanding performance requirements.

- It is also important that the selected scheduling variables reflect changes in plant dynamics as operating conditions change. It is possible in gain scheduling to incorporate linear robust control methodologies into nonlinear control design; however the global stability, robustness and performance properties are not addressed explicitly in the design process.

Though the approach is simple and the computational burden of linearization scheduling approaches is often much less than for other nonlinear design approaches, its inherent drawbacks outweigh its advantages and necessitates a new paradigm for the control of dynamical systems. New methodologies such as Adaptive control based on Artificial Neural Networks (ANN), Fuzzy logic etc try to address such problems, the lack of proof of stability and performance of such approaches over entire operating parameter regime requires design of a parameter dependent controller with guaranteed properties for which, a Linear Parameter Varying controller could be an ideal candidate.

Linear Parameter-varying Systems

LPV systems are a very special class of nonlinear systems which appears to be well suited for control of dynamical systems with parameter variations. In general, LPV techniques provide a systematic design procedure for gain-scheduled multivariable controllers. This methodology allows performance, robustness and bandwidth limitations to be incorporated into a unified framework. A brief introduction on the LPV systems and the explanation of terminologies are given below.

Parameter Dependent Systems

In control engineering, a state space representation is a mathematical model of a physical system as a set of input, u output, y and state variables, x related by first-orderdifferential equations. The dynamic evolution of a nonlinear, non-autonomous is represented by

$$\dot{x} = f(x,u,t)$$

If the system is time variant

$$\dot{x} = f(x\,(t), u\,(t), t), x(t_0)$$
$$x(t_0) = x_0, u(t_0) = u_0$$

The state variable describe the mathematical "state" of a dynamical system and in modeling large complex nonlinear systems if such state variables are chosen to be compact for the sake of practicality and simplicity, then parts of dynamic evolution of system are missing. The state space description will involve other variables called exogenous variables whose evolution is not understood or is too complicated to be modeled but affect the state variables evolution in a known manner and are measurable in real-time using sensors.

When a large number of sensors are used, some of these sensors measure outputs in the system theoretic sense as known, explicit nonlinear functions of the modeled states and time, while other sensors are accurate estimates of the exogenous variables. Hence, the model will be a time varying, nonlinear system, with the future time variation unknown, but measured by the sensors in real-time. In this case, if $w(t)$, w denotes the exogenous variable vector, and $x(t)$ denotes the modeled state, then the state equations are written as

$$\dot{x} = f(x\,(t), w(t), \dot{w}(t), u(t))$$

The parameter w is not known but its evolution is measured in real time and used for control. If the above equation of parameter dependent system is linear in time then it is called Linear Parameter Dependent systems. They are written similar to Linear Time Invariant form albeit the inclusion in time variant parameter.

$$\dot{x} = A(w(t))\,x(t) + B(w(t))\,u(t)$$
$$y = C(w(t))\,x(t) + D(w(t))\,u(t)$$

Parameter-dependent systems are linear systems, whose state-space descriptions are known functions of time-varying parameters. The time variation of each of the parameters is not known in advance, but is assumed to be measurable in real time. The controller is restricted to be a linear system, whose state-space entries depend causallyon the parameter's history. There exist three different methodologies to design a LPV controller namely,

1. Linear fractional transformations which relies on the small gain theorem for bounds on performance and robustness.
2. Single Quadratic Lyapunov Function (SQLF)
3. Parameter Dependent Quadratic Lyapunov Function (PDQLF) to bound the achievable level of performance.

These problems are solved by reformulating the control design into finite-dimensional, convex feasibility problems which can be solved exactly, and infinite-dimensional convex feasibility problems which can be solved approximately . This formulation constitutes a type of gain scheduling problem and contrast to classical gain scheduling, this approach address the effect of parameter variations with assured stability and performance.

Chapter 6

ADDITIONAL PID DESIGN AND TUNING CONCEPTS

SAMPLE TIME IS A FUNDAMENTAL DESIGN AND TUNING SPECIFICATION

There are two sample times, T, used in process controller design and tuning.

One is the control loop sample time (step 4 of the design and tuning recipe) that specifies how often the controller samples the measured process variable (PV) and computes and transmits a new controller output (CO) signal.

The other is the rate at which CO and PV data are sampled and recorded during a bump test (step 2 of the recipe).

In both cases, sampling too slow will have a negative impact on performance. Sampling fast will not necessarily provide better performance, though it may lead us to spend more than necessary on high-end instrumentation and computing resources.

Fast and slow are relative terms defined by the process time constant, Tp. Best practice for both control loopsample time and bump test data collection are the same:

Best Practice:

  Sample time should be 10 times per process time constant or faster ($T \leq 0.1Tp$).

In this article we explore both sample time issues. Specifically, we study:

1) The impact on performance when we adjust control loop sample time while keeping the tuning of a PI controller fixed, and

2) How performance is affected when we sample a process at different rates during a bump test and then complete the controller design and tuning using this same T.

The Process

Like all articles on this site, the CO and PV data we consider are the wire out to wire insamples collected at the controller interface. Thus, as shown below, the equipment (*e.g.*, actuator, valve, process unit, sensor, transmitter) and analog or digital manipulations (*e.g.*, scaling, filtering, linearization) are all lumped as a single "process" that sits between the CO and PV values.

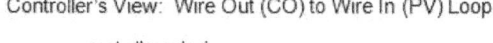

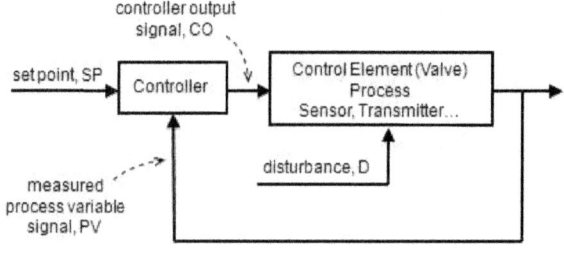

To provide the ability to manipulate and monitor all aspects of an orderly investigation, we use a differential equation (or transfer function) simulation utility to create the overall process.

It is not necessary to understand the simulation utility to appreciate and learn from the studies below. But for those interested, we provide a screen grab from the commercial software used for this purpose:

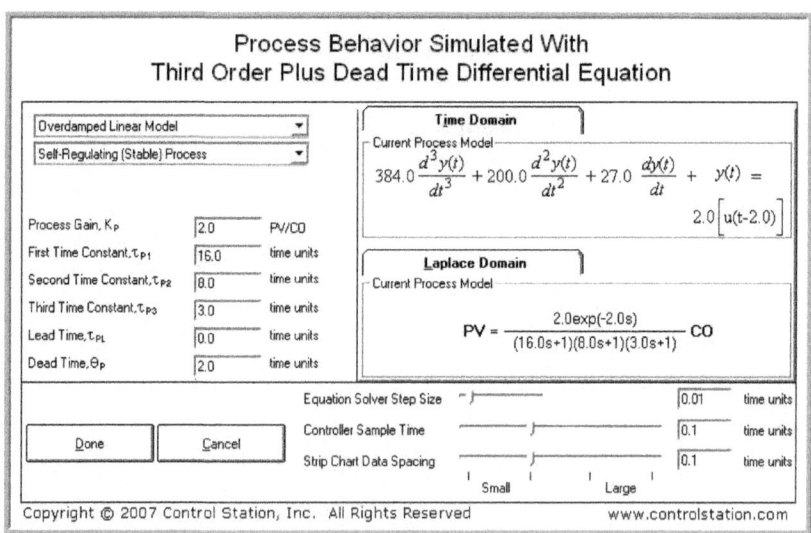

This same "process" is used in all examples in this article. Though expressed as a linear equation, it is sufficiently complex that the observations we make will be true for a broad range of process applications.

The Disturbance

Real processes have many disturbances that can disrupt operation and require corrective action from the controller. For the purposes of this investigation, we focus on only one generic disturbance (D) to our process.

The manner in which a PV responds to a disturbance is different for every application. Given this, here we choose a middle ground. Specifically, we specify that the impact of D on PV is exactly the same as the impact of CO on PV.

Thus, the D to PV behavior is simulated in the examples with the identical differential equation as shown above for the CO to PV behavior. This assumption is not right or wrong or good or bad. In fact, it is a rather common assumption in theoretical studies. Our goal is simply to provide a basis of comparison when we start exploring howsample time impacts controller performance.

The PI Controller

All examples in this article use the PI controller form:

$$CO = CO_{bias} + Kc \cdot e(t) + \frac{Kc}{Ti} \int e(t)\, dt$$

where algorithm parameters are defined here.

The PI controller computes a CO action every loop sample time, T. To best highlight differences based on sample time issues, we choose an aggressive controller tuning as detailed in this example. Tuning parameters are thus computed based on an approximating first order plus dead time (FOPDT) model fit (step 3 of the recipe) as:

$$Kc = \frac{1}{Kp} \frac{Tp}{(\theta p + Tc)} \qquad \text{and} \qquad Ti = Tp$$

where: Tc is the larger of $0.1 \cdot Tp$ or $0.8 \cdot \theta p$

Parameter Units

Time is expressed as generic "time units" and is not listed in the plots, calculations or tables. The conclusions we draw are independent of whether these are milliseconds, seconds, minutes or any other time units.

It is important to recognize, however, that when applying the observations from this article to other applications, we must be sure that all time-related parameters, including sample time, dead time, time constants and reset time, are expressed in consistent units. They should all be in seconds or all be in minutes, for example.

CO and PV both range from 0-100% in the examples. Process gain, Kp, thus has units of (% of PV)/(% of CO), while controller gain, Kc, units are (% of CO)/(% of PV).

For the remainder of this article, all parameters follow the above units convention. While something we normally avoid as bad practice, units are not explicitly displayed in the various plots and calculations.

Impact of Loop Sample Time When Controller Tuning is Constant

In this first study, we follow the recipe to design and tune a PI controller for the process above. Set point tracking performance is then explored as a function of control loop sample time, T.

Step 1: Establish the Design Level of Operation (DLO)

We arbitrarily initialize CO, PV, SP and D all at 50% and choose this as the default DLO.

Step 2: Collect CO to PV bump test data around the DLO

The plot in step 3 shows the bump test data used in this study. The sample time used during data collection is T = 0.1, which we confirm in step 3 is a very fast T = 0.005Tp.

Step 3: Approximate the Process Behavior with an FOPDT Dynamic Model

We use commercial software to fit a first order plus dead time (FOPDT) model to the dynamic process data as shown below. This fit could also be performed by hand using a graphical analysis of the plot data following the methods detailed here.

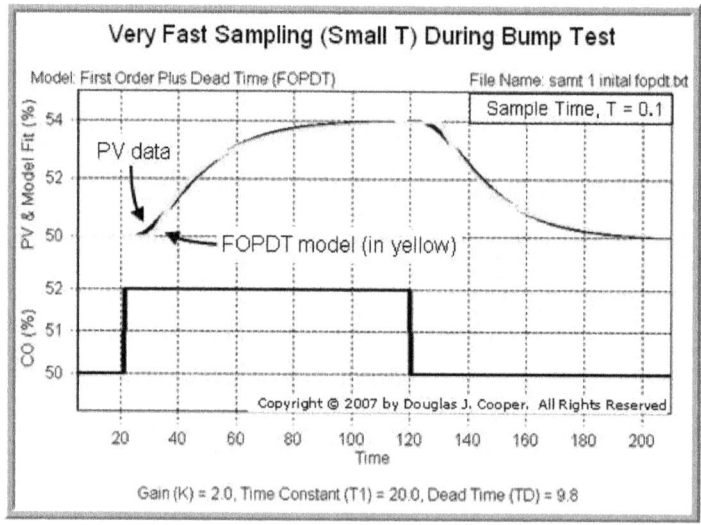

The FOPDT model fit (in yellow) visually matches the PV data in the plot, so we accept the model as an appropriate approximation of the more complex dynamic process behavior. The model parameters are listed beneath the plot and are summarized here:

- Process gain (how far), Kp = 2.0
- Time constant (how fast), Tp = 20
- Dead time (how much delay), θp = 9.8

Step 4: Use the FOPDT Model Parameters to Complete the Design and Tuning

Using the Kp, Tp and θp from step 3, we first compute the closed loop time constant, Tc, for an aggressively tuned controller as:

Aggressive Tc = the larger of 0.1 ·Tp or 0.8 θp

$$= \text{larger of } 0.1(20) \text{ or } 0.8(9.8)$$

$$= 7.8$$

and then using the tuning correlations listed earlier, we compute PI tuning values:

$$Kc = \frac{1}{2.0} \frac{20}{(9.8 + 7.8)} = 0.57 \quad \text{and} \quad Ti = 20$$

We implement these Kc and Ti values and hold them constant for the next three plots as we adjust control loopsample time, T.

The "best practice" design rule states that sample time should be 10 times per process time constant or faster ($T \leq 0.1Tp$). Hence, the maximum sample time we should consider is:

$$T_{max} = 0.1Tp = 0.1(20) = 2.0$$

The plot below shows the set point tracking performance of the PI controller with fixed tuning when the sample time T is very fast (T = 0.1), fast (T = 0.5)and right at the maximum best practice value (T = 2.0).

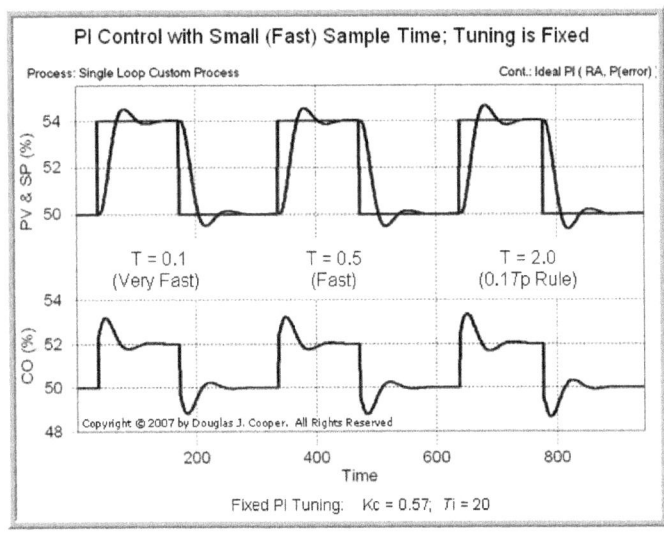

The plot above supports that a control loop sample time smaller (faster) than the "maximum best practice" 0.1 Tp rule has modest impact on performance.

In the plot below, we increase control loop sample time above the best practice limit. Because the scaling on the plot axes has changed from that above, we begin with the maximum best practice value of T = 0.1 Tp = 2.0, and then test performance when the loop sample time is slow (T = 10) and very slow (T = 20) relative to the rule.

Performance of a controller with fixed tuning clearly degrades as loop sample time, T, increases above the maximum best practice value of 0.1 Tp.

To better understand why, we zoom in on the control performance when loop sample time is very slow (T = 20). The plot below is a close-up from the plot above.

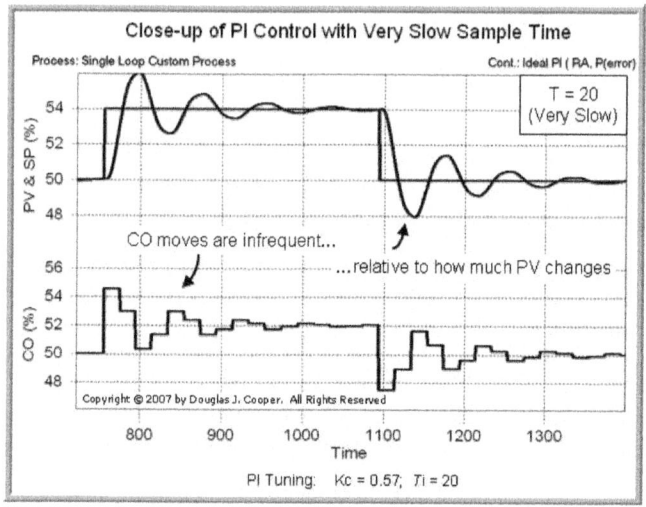

In the "very slow sample time" plot above, T = Tp. That is, the controller measures and acts only once per process time constant.

We can see in the plot that PV moves a considerable amount between each corrective CO action. At this slow sample time, the controller simply cannot keep up with the action and the result is a degraded control performance.

Impact of Data Collection Sample Time During Bump Test

In this study we sample our process at different rates, T, during a bump test, complete the design and tuning, and then test the performance of the resulting controller using this same T as the control loop sample time. Following our recipe:

Step 1: Establish the Design Level of Operation (DLO)

We again initialize CO, PV, SP and D all at 50% and choose this as the default DLO.

Step 2: Collect CO to PV bump test data around the DLO

The plots in step 3 show identical bump tests. The only difference is the sample time used to collect and record data as the PV responds to the CO steps. We consider three cases:

- sample 10 times faster than rule: $T = 0.01\,Tp = 0.2$
- sample at the maximum "best practice" rule: $T = 0.1\,Tp = 2$
- sample 10 times slower than rule: $T = Tp = 20$

Step 3: Approximate the Process Behavior with an FOPDT Dynamic Model

We again use Control Station software to fit a first order plus dead time model to each response plot (for a large view of the plots below, click $T = 0.2$; $T = 2$; $T = 20$)

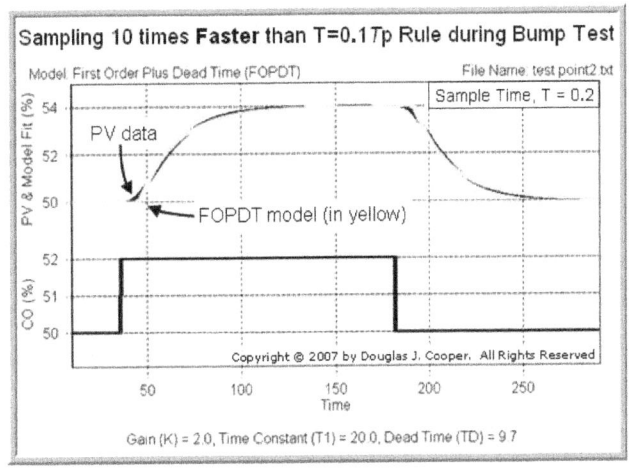

Gain (K) = 2.0, Time Constant (T1) = 20.0, Dead Time (TD) = 9.7

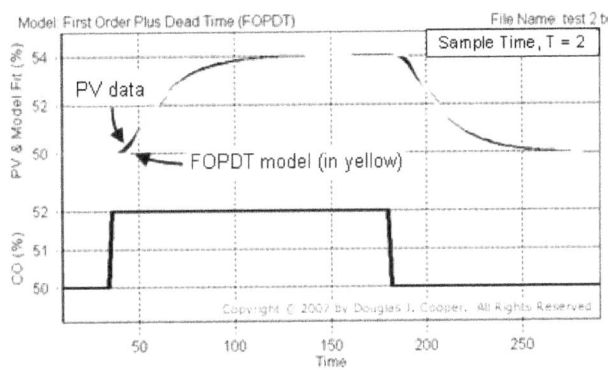

Gain (K) = 2.0, Time Constant (T1) = 20.0, Dead Time (TD) = 8.6

As the above plots reveal, an FOPDT model of the same process can be different if we sample and record data at different rates during a bump test.

Our best practice rule is based on the process time constant (Tp), a value we may not even know until after we have conducted our data collection experiment. If we do not know what to expect from a process based on prior experience, we recommend a "faster is better" attitude when sampling and recording data during a bump test.

Step 4: Use the FOPDT Model Parameters to Complete the Design and Tuning

We summarize the Kp, Tp and θp for the three cases in the table below. The first two columns ("sample fast" and "$T = 0.1\ Tp$ rule") show similar values. The "sample slow" column contains values that are clearly different.

FOPDT Model Values	Sample the **Same Process** At Different Rates During the Bump Test		
	Sample Fast $T = 0.2$	$0.1 Tp$ Rule $T = 2$	Sample Slow $T = 20$
Kp	2.0	2.0	1.9
Tp	20	20	6.0
Θp	9.7	8.6	3.2
Closed Loop Time Constant (Aggressive) Rule: Tc larger of $0.1 \cdot Tp$ or $0.8 \cdot \theta p$			Use Rule: $\theta p, min = T$
Tc	7.8	6.9	16
Aggressive PI Tuning			
Kc	0.57	0.65	0.09
Ti	20	20	6

Units: Kp [=] %PV/%CO; Tp [=] time units; θp [=] time units

Kc [=] %CO/%PV; Ti [=] time units

The tuning values shown in the bottom two rows across the table are computed by substituting the FOPDT model values into the PI tuning correlations.

The "θp, min = T" Rule for Controller Tuning

We do confront one new tuning issue in this study that merits discussion.

Consider that all controllers measure, act, then wait until next sample time; measure, act, then wait until next sample time. This "measure, act, wait" procedure has a delay (or dead time) of one sample time built naturally into its structure.

By definition, the minimum dead time, θp, in a control loop is the loop sample time, T. Dead time can certainly be larger than T (and it usually is), but it cannot be smaller.

Thus, whether by software or graphical analysis, if we compute a θp that is less than T, we must set $\theta p = T$ everywhere in our tuning correlations. This is the "θp, min = T" rule for controller tuning.

In the "sample slow" case, we know that we will be using T = 20 when we implement the controller. So even though the FOPDT model fit yields a θp = 3.2, we use θp = 20 when computing both *Tc* and *Kc* as listed in the table.

The plot below shows the set point tracking performance of three controllers, each designed and then implemented using three different sample times.

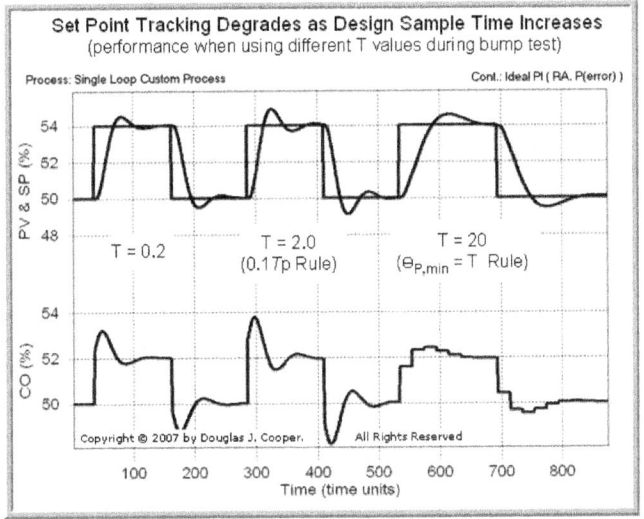

Below we show the performance of the three PI controllers in rejecting a disturbance. Recall that the D to PV dynamics are assumed to be identical to those of the CO to PV behavior.

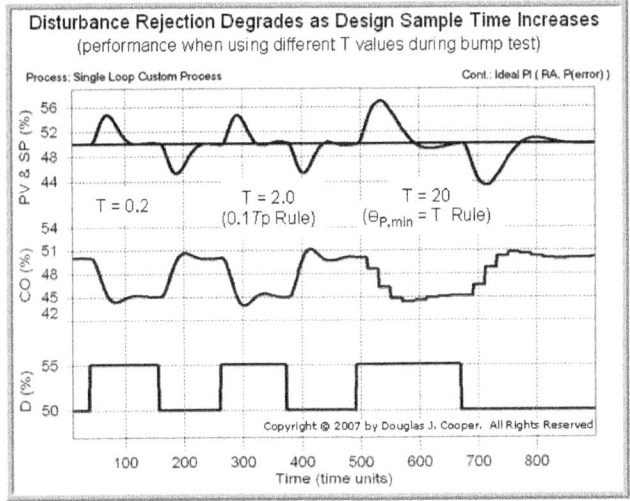

We should note that if we were to use θp = 3.2 when computing tuning values for the "sample slow" case, the controller would be wildly unstable. Even with the θp, min = T rule, performance is noticeably degraded compared to the other cases.

Final Thoughts

1) The "best practice" rule that sample time should be 10 times per process time constant or faster ($T \leq 0.1\,Tp$) provides a powerful guideline for setting an upper limit on both control loop sample time and bump test data collection sample time.

2) Sampling as slow as once per Tp during data collection and then controller implementation can produce a stable, though clearly degraded, controller. Be sure to follow the "$\theta p,min = T$" rule when using the controller tuning correlations to achieve a stable result.

PARAMETER SCHEDULING AND ADAPTIVE CONTROL OF NONLINEAR PROCESSES

Processes with streams comprised of gases, liquids, powders, slurries and melts tend to exhibit changing (or nonlinear) process behavior as operating level changes.

We discussed the nonlinear nature of the gravity drained tanks and heat exchanger processes in an earlier post. As we observed in that processes that are nonlinear with operating level will experience a degrading controller performance whenever the measured process variable (PV) moves away from the design level of operation (DLO).

We demonstrate this problem on the heat exchanger running with a PI controller tuned for a moderate response. As shown in the first set point step from 140 °C to 155 °C in the plot below, the PV responds in a manner consistent with our design goals. That is, the PV moves to the new set point (SP) in a deliberate fashion, but does not move so fast as to overshoot the set point.

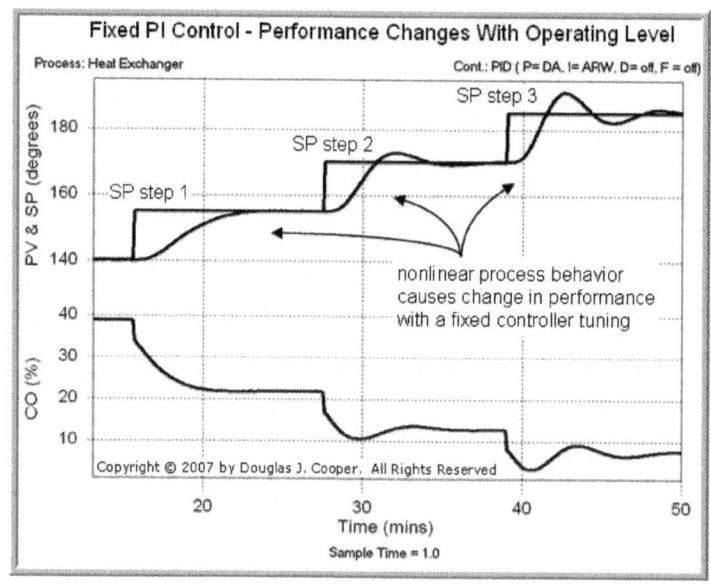

The consequence of a nonlinear process behavior is apparent as the set point steps continue to higher temperatures. In the third SP step from 170 °C to 185 °C, the same PI controller that had given a desired moderate performance now produces an active PV response with a clear overshoot and slowly damping oscillations.

If we decide that such a change in performance with operating level is not acceptable, then parameter scheduled adaptive control may be an appropriate solution.

Developing an adaptive control strategy requires additional bump tests that may disrupt production. Once sufficient process data is collected, adaptive controller design and implementation consumes more personnel time. Before we start the project, we should be sure that the loop has sufficient impact on our profitability to justify the effort and expense.

Parameter Scheduled Adaptive Control

The method of approach for parameter scheduled adaptive control is to:

a) Divide the total range of operation into some number of discrete increments or operating ranges.

b) Select a controller algorithm (P-Only, PI, PID or PID with CO Filter) for the application.

c) Specify loop sample time, action of the controller (reverse or direct acting), and other design values that will remain constant in spite of nonlinear behavior.

d) Apply our controller tuning recipe and compute tuning values for our selected controller at each of the operating increments as chosen in step a).

We require a computer based control system (a DCS or advanced PLC) to implement the adaptive logic. This is because the tuning values must be programmed as a look-up table (or schedule) where the measured PV indicates the current level of operation, and as such, "points" to appropriate controller tuning values in the table at any moment in time.

Once online, the computer reads a set of tuning values from the table as indicated by the current value of the PV. These are downloaded into the controller algorithm, which then proceeds to calculate the next controller output (CO) value.

Tuning updates are downloaded into the controller every loop sample time, T. As a result, the controller continually adapts as the operating level changes to maintain a reasonably consistent control performance across a range of nonlinear behavior.

Notes:

1) The set point (SP) is not appropriate to use as the operating level "pointer" because it indicates where we hope the PV will be, not necessarily where it

actually is. The CO value can change both as operating level changes and as the controller works to reject disturbances. Since it reflects both disturbance load and operating level, the correspondence between current CO and current operating level is inconsistent. Current PV offers the most reliable indicator of expected process behavior.

2) "Gain scheduling" is a simplified variation of parameter scheduling, where, rather than updating all tuning values as operating level changes, only the controller gain is updated. All other tuning values remain constant with a pure gain scheduling approach. This simplification increases the chance that important process behaviors (such as a changing dead time, θp) will be overlooked, thus decreasing the potential benefit of the adaptive strategy. With modern computing capability now widely available, we see no benefit from a "gain only" simplification unless it is the only choice offered by our vendor.

Interpolation Saves Time and Money

Ultimately, it is impractical to divide our range of operation into many increments and then tune a controller at each level. Such an approach requires that we bump the process at least once in each operating increment. As we had alluded to in previous discussion, this can cause significant disruption to the production schedule, increase waste generation, use expensive materials and utilities, consume precious personnel time, and everything else that makes any loop tuning project difficult to sell in a production environment.

Thus, a popular variation on parameter scheduling, and the one explored in the case study below, is to design and tune only three controllers that span the range of expected operation. We then interpolate (fit a line) between the tuning values so we can update (or adapt) our controller to match any operating level at any time.

Normally, one controller is tuned to operate near the lower set point value we expect to encounter, while another is tuned for the expected high SP value. The third controller is then tuned for a strategically located mid range operation to give an appropriate shape to our interpolation curve. Our goal in choosing where to locate this midpoint is to reasonably approximate the complex nonlinear behavior of our process while keeping disruptive testing to a minimum. More discussion follows in the case study.

Case Study: Adaptive Control of the Heat Exchanger

We use the heat exchanger to illustrate and explore the ideas introduced above. As always, we follow our controller design and tuning recipe as we proceed. We choose a PI controller for the study and use the constant design values as detailed here (*e.g.*, dependent PI algorithm form; loop sample time, $T = 1.0$ sec; controller is direct acting).

Step 1: Design Level of Operation (DLO)

Our adaptive schedule requires tuning values for three PI controllers that span the range of expected operation. Hence, we need to specify three design levels of operation (DLOs), one for each controller.

Since, as we detail here, set point driven data can be analyzed with commercial software for controller design and tuning, we will use the data from the plot at the top of this post as our bump test data.

A good bump test should generate dynamic process data both above and below the DLO. This is best practice because we then "average out" nonlinear effects when, in step 3 that follows, we approximate the bump test data with a simplifying first order plus dead time (FOPDT) dynamic model.

Since our test data has already been collected and plotted, we reverse this logic and pick the three DLOs as the midpoint of each SP step. Reading from the plot as shown below, we thus arrive at: DLO 1 = 147 °C; DLO 2 = 163 °C; DLO 3 = 178 °C

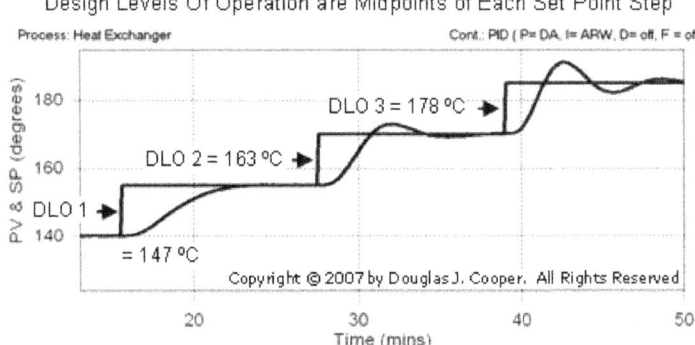

Step 2: Collect Process Data around the DLO

The plot data provides us with three complete CO to PV bump tests, each centered around its DLO. Hence, step 2 is complete.

Step 3: Fit a FOPDT Model to the Dynamic Process Data

We use Control Station's Loop-Pro software to divide the plot data into three CO to PV bump tests. We then use the software to fit a FOPDT model to each bump test following the same procedure as detailed here.

The plots below show the data and FOPDT model approximations. Because each FOPDT model visually matches its bump test data, we have confidence that the model parameters reasonably describe the dynamic process behavior at the three design levels of operation.

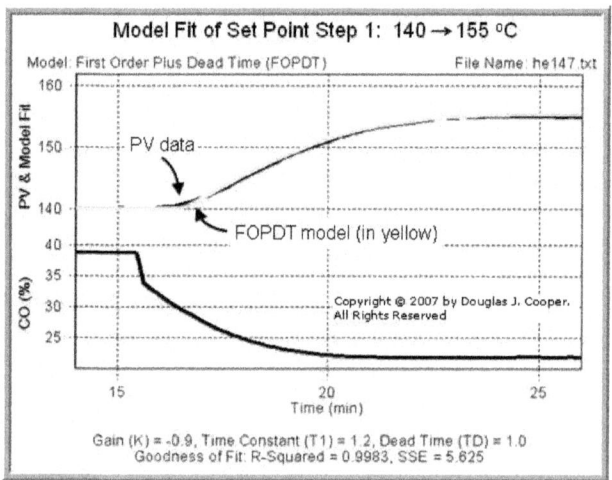

Gain (K) = -0.9, Time Constant (T1) = 1.2, Dead Time (TD) = 1.0
Goodness of Fit: R-Squared = 0.9983, SSE = 5.625

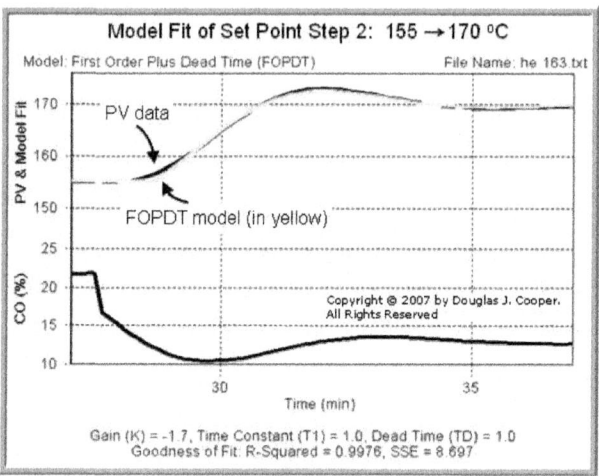

Gain (K) = -1.7, Time Constant (T1) = 1.0, Dead Time (TD) = 1.0
Goodness of Fit: R-Squared = 0.9976, SSE = 8.697

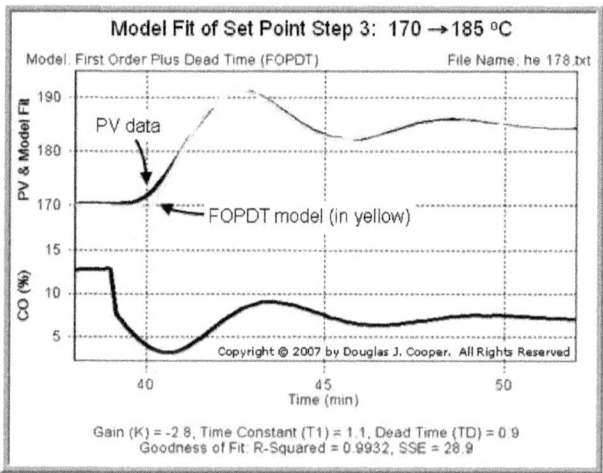

Gain (K) = -2.8, Time Constant (T1) = 1.1, Dead Time (TD) = 0.9
Goodness of Fit: R-Squared = 0.9932, SSE = 28.9

Step 4: Use the FOPDT Model Parameters to Complete the Design

The table below summarizes the FOPDT model parameters from step 3 for the three DLOs.

FOPDT Model Values	SP Step 1 140 →155 °C DLO 1 = 147 °C	SP Step 2 155 →170 °C DLO 2 =163 °C	SP Step 3 170 →185 °C DLO 3 =178 °C
Kp	− 0.9	− 1.7	− 2.8
Tp	1.2	1.0	1.1
Θp	1.0	1.0	0.9
Moderate PI Tuning			
Kc	− 0.15	− 0.07	− 0.05
Ti	1.2	1.0	1.1
Aggressive PI Tuning			
Kc	− 0.74	− 0.33	− 0.24
Ti	1.2	1.0	1.1

Units: Kp [=] °C/%, Tp [=] min, Θp [=] min

Kc [=] %/°C, Ti [=] min

Note that the process gain, Kp, varies by 300% (from -0.9 to -2.8 °C/%) across the operating range. In contrast, process time constant, Tp, and process dead time, θp, change by quite modest amounts.

For this investigation, we compute both moderate and aggressive PI tuning values for each DLO. The rules and correlations for this are detailed in this post, but briefly, we compute our closed loop time constant, Tc, as:

aggressive: Tc is the larger of $0.1 \cdot Tp$ or $0.8 \, \theta p$

moderate: Tc is the larger of $1 \cdot Tp$ or $8 \, \theta p$

With Tc computed, the PI correlations for controller gain, Kc, and reset time, Ti, are:

$$Kc = \frac{1}{Kp} \frac{Tp}{(\theta p + Tc)} \quad \text{and} \quad Ti = Tp$$

The moderate and aggressive PI tuning values for each of the DLOs are also summarized in the table above.

Below we illustrate how to interpolate controller gain, Kc, for the moderate tuning case.

As shown in the plot, the three moderate Kc values are plotted as a function of PV. Lines of interpolation are fitted between each Kc value. The equations for these lines must be programmed into the control computer.

Now, as PV moves anywhere from 147 to 178 °C, we can use these equations to compute a unique value for Kc. This concept must also be applied to obtain interpolating equations for reset time, Ti, thus producing a fully adaptive controller.

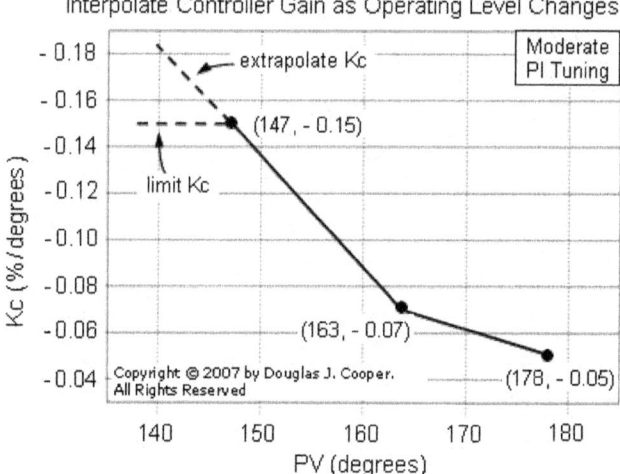

As shown in the plot above, one decision that must be made is whether to extrapolate the line and have the parameter continue the trend past the actual maximum or minimum data point. Alternatively, we could choose to limit Kc and have it stay constant for all PV values beyond the maximum or minimum.

Unless we are confident that we understand the true nature of a process, extrapolation into the unknown is more often a bad idea than a good one. In this case study, we choose not to extrapolate. Rather, we limit the tuning parameters to the maximum and minimum DLO values in the table. That is, Kc remains constant at -0.15 %/°C when PV moves below 147 °C, and remains constant at -0.05 when Kc moves above 178 °C. In between, Kc tracks the interpolating lines in the plot above.

The capability of this parameter scheduled adaptive control is shown for a moderate PI controller in the plot below. To appreciate the difference, compare this constant performance to the varied response in the plot at the top of this post.

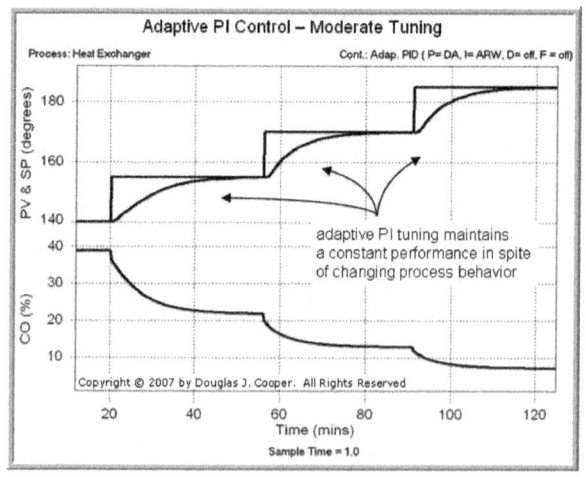

The result of an aggressively tuned PI controller is shown in the plot below. The performance response is again quite consistent (though admittedly not perfect) across the range of operation.

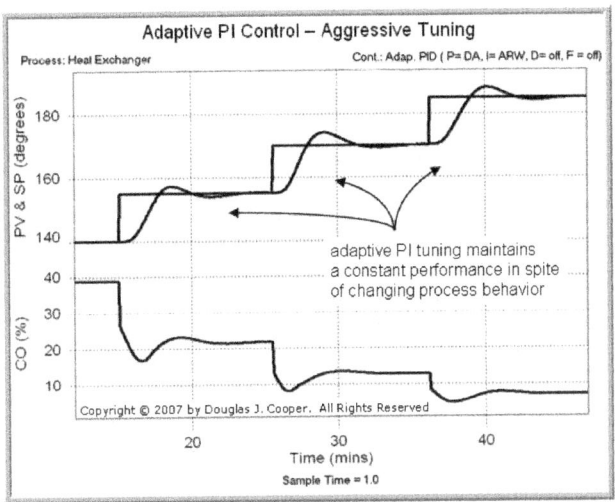

A Proven Strategy

As these plots illustrate, a parameter scheduled adaptive controller can achieve consistent performance on processes that are nonlinear with operating level. This adaptive strategy has been widely employed in industrial practice and, as shown in this case study, is quite powerful in addressing a challenging and important problem.

Again, however, design and implementation requires extra effort and expense. We should be sure the loop is important enough to warrant such an investment before we begin.

PLANT-WIDE CONTROL REQUIRES A STRONG PID FOUNDATION

The term "plant-wide control" is used here to describe the use of advanced software that sits above (or on top of) the individual PID controllers running a number of process units in a plant.

Depending on the technology employed, this advanced process control software can perform a variety of predictive, scheduling, supervisory and/or optimizing computations. The most common architecture in industrial practice has the plant-wide software compute and transmit set point updates to a traditional platform of individual PID controllers.

Notes:

- Plant-wide control, where one package computes updates for an entire facility.
- While it is possible for an advanced control package to completely eliminate the need for individual PID controllers, this is also a rare practice. One reason

is that advanced software is often an add-on to a plant already in operation. The existing PID controllers provide a distributed backup infrastructure that enable continued operation, including an orderly shutdown, in the event of a computer problem. Even in new construction, plants are normally built with a traditional platform of PID controllers underneath the advanced control software.

Plant-wide process control holds great allure. The promise is an orchestrated operation for addressing upsets, maximizing throughput, minimizing energy and environmental impact, making scheduling more flexible, and keeping production to a tighter specification.

The figure below illustrates the hierarchy of a more complex multi-level implementation.

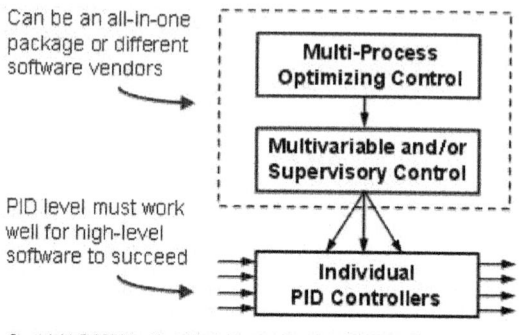

Higher Level Software Uses Longer Time Scales

The individual PID loops of a traditional control system provide the firm foundation for advanced process control. Typical loop sample times for PID controllers on processes with streams comprised of gases, liquids, powders, slurries and melts are often on the order of once per second.

Advanced software sitting above the individual controllers computes and transmits values less frequently, perhaps on the order of once every ten seconds to several minutes. High level optimizers output commands even less often, ranging from once per hour to even once per day. The frequency depends on the time constants of the process units in the hierarchy, the complexity of the plant-level control objectives, the numerical solution methods employed, and the capabilities of the installed hardware.

Project Stages For a Retrofit

Though each implementation is different, projects on existing plants tend to follow a standard progression:

1) Validate all sensors and associated instrumentation; replace where necessary.

2) Service all valves, pumps, compressors and other final control elements to ensure proper function. Replace or upgrade where necessary.

3) Upgrade the DCS (distributed control system) with latest software releases; update hardware if it is aging.

4) Tune all low level PID loops, including cascade, ratio and similar architectures.

5) Upgrade the computers in the control room to modern standards; ensure enough computing power to handle the plant-level software.

6) Design and deploy the plant-level control software.

Note that step 6 is presented as a simplistic single step. In reality, the design and implementation of plant-level software is a complex procedure requiring tremendous experience and sophistication on the part of the project team. Step 6 is presented as a summary bullet item because the purpose of this post is to separate and highlight the vital importance of a properly operating platform of individual PID controllers in any control project.

Steps 1-4 Provide Profitability

Software vendors who suggest that a prospective company consider all steps 1-6 as part of an "advanced process control" project are not being completely transparent.

The return on investment (ROI) may appear attractive for the complete project, but it is appropriate to determine what portion of the return is provided by steps 1-4 alone. The profit potential of these first steps, reasonably characterized as traditional control tasks, can be responsible for well over half of the entire revenue benefit on some projects!

Arguably, it is a better business practice to work through steps 1-4 and then reevaluate the situation before making a decision about the need for and profit potential from plant-level control software.

When the base level instrumentation is in proper working order, the PID loops can be tuned to provide improved plant performance. Rather than using historical tunings, the project team should use commercial software to quickly analyze process data and compute appropriate tuning parameters.

PID Control is The Foundation

As the figure at the top of this post illustrates, the PID loops provide the strong foundation upon which the plant-level software sits.

Plant-wide process control software cannot hope to improve plant operation if the PID loops it is orchestrating are not functioning properly. It is folly to proceed with step 6 above before having worked through steps 1 – 4.

In a great many situations, by the time step 4 is completed, indeed, the plant will be running significantly better. The orderly maintenance and tuning of the first level PID loops will provide a fast payback and begin making money for the company. Quickly.

ZIEGLER-NICHOLS TUNING POOR CHOICE FOR PRODUCTION PROCESSES

Ziegler and Nichols first proposed their method in 1942. It is a trial-and-error loop tuning technique that is still widely used today. The automatic mode (closed-loop) procedure is as follows:

- Set our controller to P-Only action and switch it to automatic when the process is at the design level of operation.

- Guess an initial controller gain, Kc, that we expect (hope) is conservative enough to keep the loop stable.

- Bump the set point a small amount and observe the response behavior.

- If the controller is not causing the measured process variable (PV) to sustain an oscillating pattern, increase the Kc (or decrease the proportional band, PB) and wait for the new response.

- Keep adjusting and observing by trail and error until we discover the value of Kc that causes sustained uniform oscillations in the PV. These oscillations should neither be growing nor dying out, and the controller output (CO) should remain unconstrained.

- The controller gain at this condition is called the *ultimate gain*, Ku. The period of the PV oscillation pattern at the ultimate gain is called the *ultimate period*, Pu.

After using the procedure above to determine the ultimate gain and ultimate period, the Ziegler-Nichols (Z-N) correlations to compute our final tuning for a PI controller are:

- Ziegler-Nichols PI Tuning: Kc = 0.45·Ku   Ti = Pu/1.2

Many process control books and articles propose a variety of tuning correlations using Ku and Pu that provide improved performance over that provided by the original Z-N correlations listed above. Of course, the definition of "improved" can vary widely depending on the process being controlled and the operations staff responsible for a safe and profitable operation.

Ziegler-Nichols Applied to the Heat Exchanger

To illustrate the procedure, we apply it to the same heat exchanger process explored in numerous articles in this e-book.

Below is a plot showing the heat exchanger under P-Only control. The controller gain, Kc, is initially set to Kc = −1.0 %/°C, a value we hope provides conservative performance as we start our guess and test investigation.

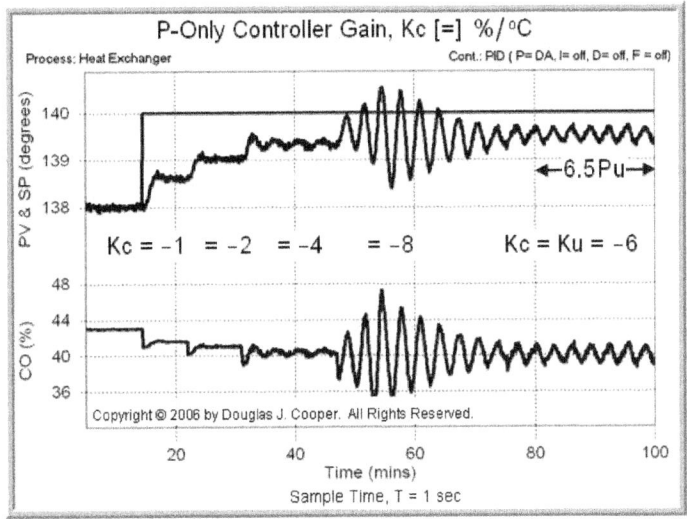

As shown above, the process is perturbed with a set point step and we wait to see if the controller yields sustained oscillations. The process variable (PV) displays a sluggish response at Kc = -1.0 %/°C, so we double the controller gain to Kc = -2.0 %/°C, and then double it again to Kc = -4.0 %/°C, waiting each time for the response pattern to establish itself.

When changing from a Kc = -4.0 up to Kc = -8.0 %/°C as shown in the above plot at about time t = 45 min, the heat exchanger goes unstable as evidenced by the rapidly growing oscillations in the PV. This indicates we have gone too far and must back down on the P-Only controller gain.

A compromise of Kc = -6.0 %/°C seems to create our desired goal of a process teetering on the brink of stability. Hence, we record our ultimate gain as:

$$Ku = -6 \ \%/°C$$

As indicated in the plot above from time t = 80 min through t = 100 min, when under P-Only control at our ultimate gain, the PV experiences 6.5 complete cycles or periods of oscillation over 20 minutes. The ultimate period, Pu, is thus computed as:

$$Pu = 20 \ min/6.5 = 3.1 \ min$$

We employ these Ku and Pu values in the Z-N correlations above to obtain our PI tuning values:

- Kc = 0.45·Ku = 0.45 (-6 %/°C) = -2.7 %/°C
- Ti = Pu/1.2 = 3.1 min/1.2 = 2.6 min

Below (click for a large view) is the performance of this Z-N tuned PI controller in tracking a set point step from 138 °C up to 140 °C. For comparison, the heat exchanger responding to the same set point steps using an IMC tuned PI controller is shown here.

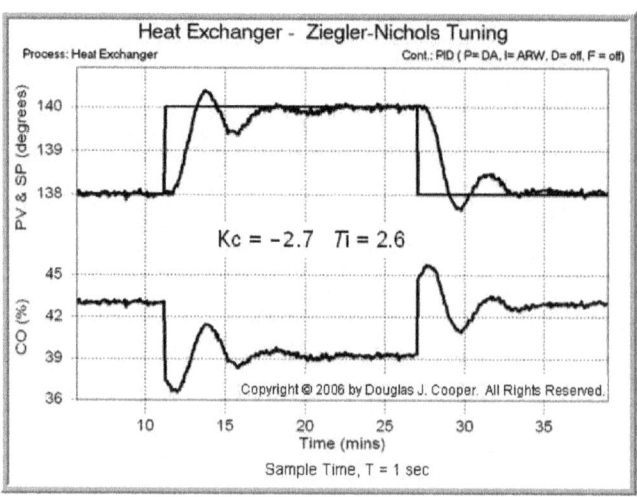

Concerns in a Production Environment

Suppose we were using the Z-N closed-loop method to tune the above controller in a production environment:

1. The trial and error search for the ultimate gain took about 80 minutes in the test above. For comparison, consider that the single step test for the IMC design on the heat exchanger took only 5 minutes.

Even if there is suspicion that we skewed the process testing to make Z-N look bad, the fact remains that it will always take significantly longer to find the ultimate gain by trial and error then it will to perform a simple bump test.

And in a production situation, this means we are losing time and wasting money as we make bad product. Why even consider such a time-consuming and wasteful tuning method when it is not necessary?

2. Perhaps more significant, the Z-N method requires that we literally bring our process to the brink of instability as we search for the ultimate gain. Creeping up on the ultimate gain can be very time consuming (and hence very expensive), but if we try to save time by making large adjustments in our search for Ku, it becomes much more likely that we will actually go unstable, at least for a brief period.

In production situations, especially involving reactors, separators, furnaces, columns and such, approaching an unstable operation is an alarming proposition for plant personnel.

Final Thoughts

This e-book focuses on the control of processes with streams composed of liquids, gases, powders, slurries and melts. The challenging control problems for such process are most always found in a production environment.

Our controller design and tuning recipe is designed specifically for such situations. It's fast and efficient, saves time and money, and provides consistent and predictable performance.

CONTROLLER TUNING USING SET POINT DRIVEN DATA

The controller design and tuning recipe we have used so successfully on this site requires that we bump our process and collect dynamic data as the process responds.

For the heat exchanger and gravity drained tanks study, we generated dynamic data using a step test while the controller was in manual mode. One benefit of this "open loop" step test is that we can analyze the response graph by hand to compute first order plus dead time (FOPDT) dynamic model parameters. These parameter values are then used in rules and correlations to complete the controller design and tuning.

In a production environment, operations personnel may not be willing to open a loop and switch to manual mode "just" so we can perform a bump test on a process. In these situations, we must be prepared to perform our dynamic testing when the controller is in automatic.

Arguably, such closed-loop testing offers potential benefits. Presumably, operation is safer while under automatic control. And closed loop testing should enable us to generate dynamic data and return to steady operation faster than other bump test method.

Software Required

Unfortunately, once we deviate from the pure manual mode step test, we must use a software tool to fit an FOPDT model to the data. Fortunately, inexpensive software is available that helps us perform our data analysis andcontroller tuning chores quickly and reliably.

In theory, closed loop dynamic testing can be problematic because the information contained in the process data will reflect both the character of the controller as well as that of the process. If we remain conscious of this and take modest precautions, this theoretical concern rarely causes real world problems.

Generating Dynamic Data in Automatic Mode

For closed loop studies, dynamic test data is generated by stepping, pulsing or otherwise bumping the set point.

For model fitting purposes, the controller, when working to track this set point bump, must take actions that are energetic enough to generate a proper data set, but not be so aggressive that the PV oscillates wildly during data collection.

Similar to manual mode testing, a proper data set is generated when the controller output (CO) is moved far enough and fast enough to force a clear response in the measured process variable (PV).

Also, similar to manual mode testing, the process must be at a steady operation before first bumping the set point. The point of a bump test is to learn about the cause-and-effect relationship between the CO and PV. With the plant at a steady operation, we are starting with a clean slate and the dynamic character of the process will be clearly revealed as the PV responds. Just be sure to have the data capture routine collect the entire event, starting from before the initial bump.

Set Point Bumps to the Heat Exchanger

Below is the heat exchanger process under P-Only control. As shown, the process is initially at a steady operation. The set point is stepped from 138 °C up to 140 °C and back again. The P-Only controller produces a moderate set point response, with offset displayed as expected from this simple controller form.

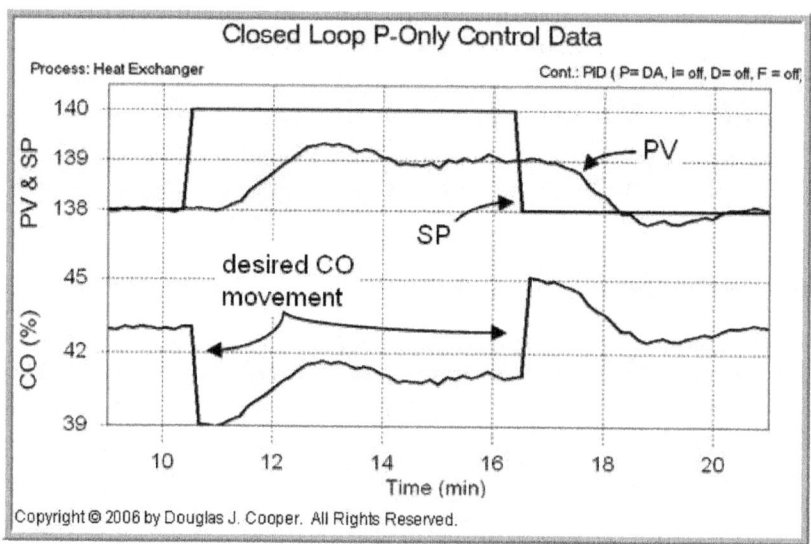

What is important in the above test is that when the set point is stepped, the P-Only controller is active enough to move the CO in the desired "far enough and fast enough" manner to force a clear response in the PV trace. This obvious CO to PV cause-and-effect relationship is exactly what we require in a good data set.

Below is the heat exchanger process with a poorly tuned PI controller. Again, the process is initially at a steady operation and the set point is stepped from 138 °C up to 140 °C and back.

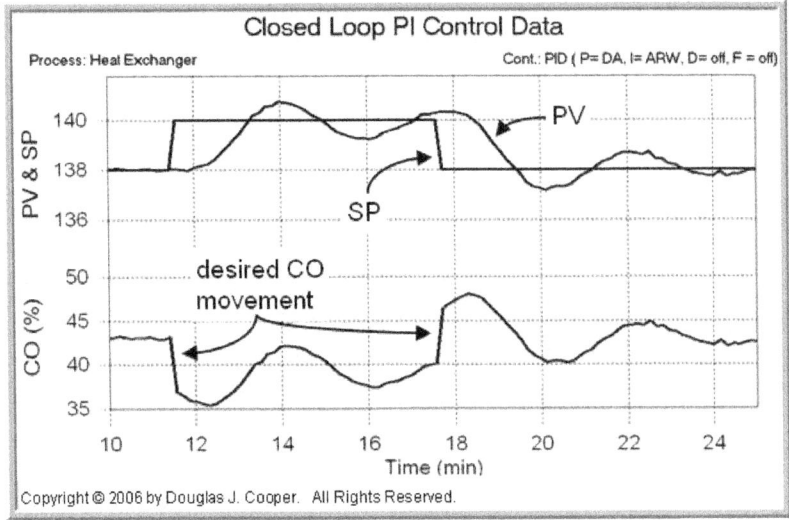

This process is approaching the upper extreme of energetic oscillations. But as noted on the plot, we can still see the desired clear and sudden CO movement that forces the PV dynamics needed for a good FOPDT model fit.

Automated Model Fitting

Below we fit a FOPDT model to the process data from the above two set point bump tests. We use the Control Station *Engineer* software in these examples.

Here is an automated model fit of the above P-Only control data using the Control Station software:

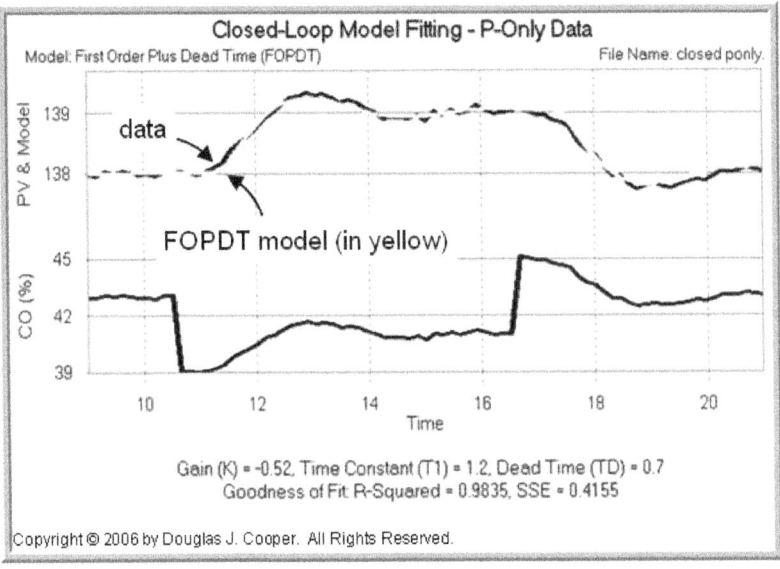

And here is an automated model fit of the above PI control data using the software:

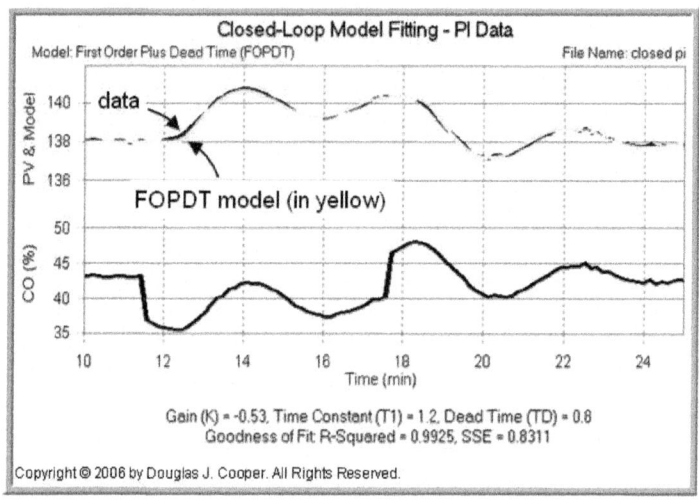

Gain (K) = -0.53, Time Constant (T1) = 1.2, Dead Time (TD) = 0.8
Goodness of Fit: R-Squared = 0.9925, SSE = 0.8311

Comparing Results

The table below compares the FOPDT model parameters from the above set point driven bump test against the open loop step test results presented earlier:

	Open Loop Step Test	Closed Loop P-Only Data	Closed Loop PI Data
Process Gain, Kp	-0.53	-0.52	-0.53
Time Constant, Tp	1.3	1.2	1.2
Dead Time, Θp	0.8	0.7	0.8

The table confirms what we can see visually in the model fit plots; it is certainly possible to obtain an accurate FOPDT dynamic model from closed loop data.

Implications for Control

The small differences in parameter values in the above table will have negligible impact on controller design and tuning. In fact, real world process data is imperfect and differences should be expected, even when repeating the identical test on the same process.

DO NOT MODEL DISTURBANCE DRIVEN DATA FOR CONTROLLER TUNING

A fairly common stumbling block for those new to controller tuning relates to step 2 of the controller design and tuning recipe. Step 2 says to "collect controller output (CO) to process variable (PV) dynamic process data around the design level of operation."

But suppose disturbance rejection is our primary control objective (example study here). Shouldn't we then step or pulse (or "bump") our disturbance variable to generate the step 2 dynamic process test data?

As shown below for the gravity drained tanks process, that would involve bumping D, the flow rate of the pumped stream exiting the bottom tank and modeling the D to PV dynamic relationship:

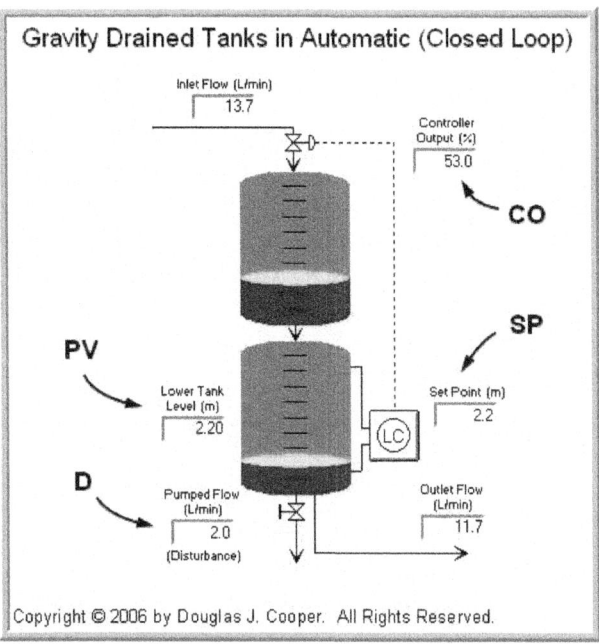

The short answer is, no. Tuning a feedback controller based on the D to PV dynamic behavior is a path to certain failure.

Wire Out to Wire In

A controller's "world" is wire out to wire in. The CO is the only thing a controller can adjust. The PV is the only thing a controller can "see." The controller sends a CO signal out on one wire. The impact of that action returns as a PV measurement signal on the other wire.

Disturbances, by their very nature, are often unmeasured. Unless a feed forward architecture has been implemented, the controller is only aware of a disturbance when it has already forced the PV from set point (SP). The CO is then the only handle the controller has to correct the problem.

For a controller to take appropriate corrective actions, it must "know" how the PV will respond when it changes the CO. Thus, regardless of the control objective, it is CO to PV relationship that must always be the foundation for controller design and tuning.

Closed Loop Testing

As discussed here, we must use a software tool to fit a model to dynamic process test data that has been collected in automatic mode.

As always, the process must be steady before beginning the test. Also, the controller must be tuned so that the CO actions are energetic enough to force a clear response in the PV, but not be so aggressive that the PV oscillates wildly during data collection.

SP Driven Dynamic Data is Good

Useful data can be generated by bumping the set point enough to force a clear dynamic response. Below is data from the gravity drained tanks process under P-Only control using a controller gain, Kc = 16 %/m. As shown, the process is initially at a steady operation. The set point is stepped in a doublet, from 2.2 m up to 2.4 m, then down to 2.0 m, and back to the initial 2.2 m. The P-Only controller produces a moderate set point response, with offset displayed as expected from this simple controller. While not shown, the pumped flow disturbance, D, remains constant throughout the experiment.

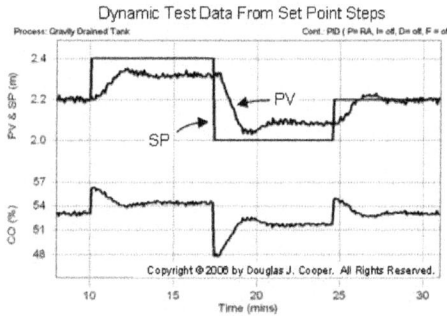

What is important in the above test is that when the set point is stepped, the P-Only controller is sufficiently active to move the CO in the desired "far enough and fast enough" manner to force a clear response in the PV trace. This obvious CO to PV cause-and-effect relationship is exactly what we require in a good data set.

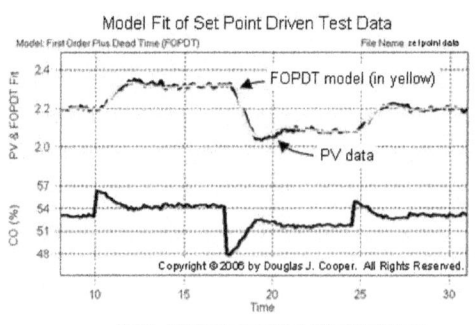

Step 3 of the controller design and tuning recipe is to fit a FOPDT (first order plus dead time) model to the dynamic process test data. Below is the results of this fit using the above set point driven test data and the Control Station software.

These results will be discussed later in this article.

D Driven Dynamic Data is NOT Good

Next we conduct a dynamic test where the set point remains constant and the dynamic event is forced by changes in D, the pumped flow disturbance.

As shown below, D is stepped from 2 L/min up to 3 L/min, down to 1 L/min, and back to 2 L/min. The same P-Only controller used above produces a moderate disturbance rejection response with offset (more discussion on P-Only control, disturbance rejection and offset for the gravity drained tanks can be <u>found here</u>).

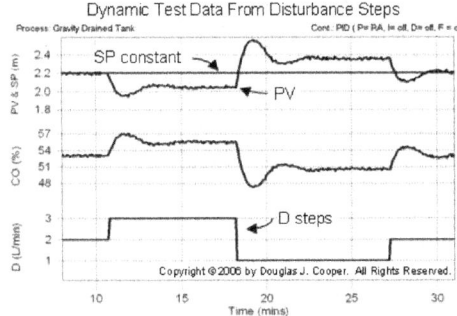

As per step 3 of the recipe, shown below is a FOPDT model fit of the dynamic process test data from the above experiment:

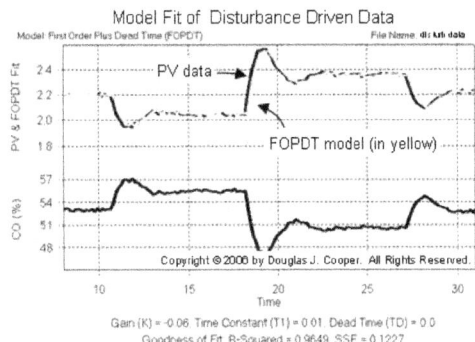

It is unfortunate that the model fit looks so good, because it may give us confidence that the design is proceeding correctly.

Comparing Model Fit Results

The table below summarizes our FOPDT model parameters resulting from:

* an open loop step test as described here,

- the closed loop set point driven test shown above,
- the closed loop disturbance driven test shown above.

	Open Loop Step	*--- closed loop tests ---* Set Point Driven	Disturbance Driven
Process gain, Kp (m/%)	0.09	0.09	-0.06
Time constant, 7p (min)	1.4	1.4	0.01
Dead time, Θp (min)	0.5	0.4	0.0

As expected, the set point driven test produces model parameters that are virtually identical to those of the open loop test. And therefore, the controller design and tuning based on either of these two tests will provide the same desirable performance (PI control study using these parameters presented here).

But the disturbance driven model is distressing. The modeling fitting software succeeds in accurately describing the data (and this is a wonderful capability for feed forward control element design), but the parameters of the disturbance driven model are very different from those needed for proper control of the gravity drained tanks.

Perhaps most striking is that the disturbance driven test data yields a negative process gain, Kp. A controller designed from this data would have the wrong action (direct acting or reverse acting). And as a result, the controller would move the valve in the wrong direction, compounding errors rather than correcting for them.

Disturbances are Always Bad

When generating dynamic process test data, it is essential that the influential disturbances remain quiet. If we are not familiar enough with our process to be sure about such disturbances, it would be best that we not adjust any controller settings.

COMPARING CONTROLLER PERFORMANCE USING PLOT DATA

When considering the range of control challenges found across the process industries, it becomes apparent that very different controller behaviors can be considered "good" performance. While one process may be best operated with a fast and aggressive control action, another may be better suited for a slow and gentle response.

Since there is no common definition of what is good or best performance, we are the judge and jury of goodness for our own process.

Performance is a Matter of Application

Suppose our process throughput is known to change quite suddenly because of the unreliable nature of our product filling/packaging stations at the end of our production line. When one of the container filling stations goes down, the upstream process throughput must be ramped down quickly to compensate.

And as soon as the problem is corrected, we seek to return to full production as rapidly as possible.

In this application, we may choose to tune our controllers to respond aggressively to sudden changes in throughput demand. We must recognize and accept that the consequence of such aggressive action is that our process variables (PVs) may overshoot and oscillate as they settle out after each event.

Now suppose we work for a bio-pharma company where we grow live cell cultures in large bioreactors. Cells do not do well and can even die when conditions change too quickly. To avoid stressing the culture, it is critical that the controllers move the process variables slowly when they are counteracting disturbances or working to track set point changes.

So good performance can sometimes mean fast and aggressive, and other times slow and gentle.

Sometimes Performance is a Matter of Location

Distillation columns are towers of steel that can be as tall as a 20 story building. They are designed to separate mixtures of liquids into heavier and lighter components. To achieve this, they must run at precise temperatures, pressures and flow rates.

Because of their massive size, they have time constants measured in hours. A disturbance upset in one of the these behemoths can cause fluctuations (called "swings") in conditions that can take the better part of a day to settle out. If the plant is highly integrated with streams coming and going from process units scattered across the facility, then a distillation column swinging back and forth for hours will cause the operation of units throughout the plant to be disrupted.

So the rule of thumb in such an integrated operation is that the columns are king. If we are involved in operating a process upstream of a distillation column, we do everything in our power to contain all disruptions within our unit. Good control means keeping our problems to ourselves and avoiding actions that will impact the columns. We do this no matter what havoc it creates for our own production.

And if we are involved in operating a process downstream from a column, our life can be similarly miserable. The column controllers are permitted to take whatever action is necessary to keep the columns running smoothly. And this means that streams flowing out of the column can change drastically from moment to moment as the controllers fight to maintain balance.

A Performance Checklist

As these examples illustrate, good control performance is application specific. Ultimately, we define "best" based on our own knowledge and experience. Things we should take into account as we form our judgments include what our process is physically able to achieve, how this fits into the bigger safety and profitability picture of the facility, and what management has planned.

Thus, our performance checklist requires consideration of the:

- goals of production
- capabilities of the process
- impact on down stream units
- desires of management

Controller Performance from Plot Data

If we limit ourselves to judging the performance of a specific controller on one of our own processes, we can compare response plots side-by-side and make meaningful statements about which result is "better" as we define it.

There are more precise terms beyond "oscillates a lot" or "responds quickly." We explore below several criteria that are computed directly from plot data. The more common terms in this category include:

- rise time
- peak overshoot ratio
- settling time
- decay ratio

These and like terms permit us to make orderly comparisons among the range of performance available to us.

Peak Related Criteria

Below is a set point step response plot with labels indicating peak features:

A = size of the set point step

B = height of the first peak

C = height of the second peak

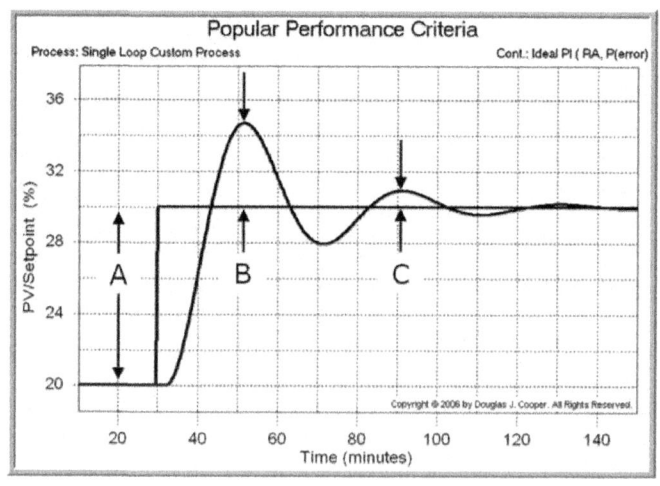

The popular peak related criteria include:

- Peak Overshoot Ratio (POR) = B/A
- Decay Ratio = C/B

In the plot above, the PV was initially at 20% and a set point step moves it to 30%. Applying the peak related criteria by reading off the PV axis:

A = (30 – 20) = 10%

B = (34.5 – 30) = 4.5%

C = (31 – 30) = 1%

And so for this response:

POR = 4.5/10 = 0.45 or 45%

Decay ratio = 1/4.5 = 0.22 or 22%

An old rule of thumb is that a 10% POR and 25% decay ratio (sometimes called a quarter decay) are popular values. Yet in today's industrial practice, many plants require a "fast response but no overshoot" control performance.

No overshoot means no peaks, and thus, B = C = 0. This increasingly common definition of "good" performance means the peak related criteria discussed above are not useful or sufficient as performance comparison measures.

Time Related Criteria

An additional set of measures focus on time-related criteria. Below is the same set point response plot but with the time of certain events labeled :

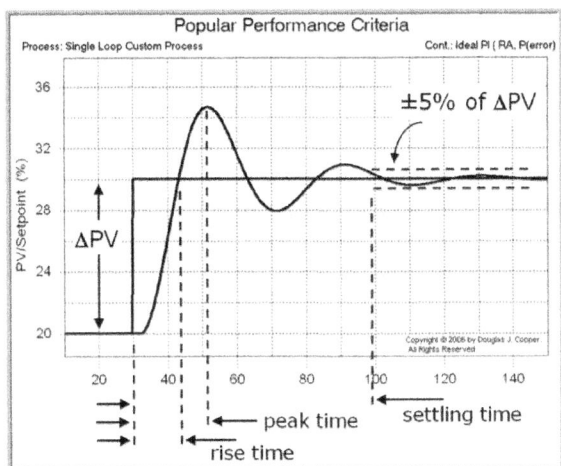

The clock for time related events begins when the set point is stepped, and as shown in the plot, include:

- Rise Time = time until the PV first crosses the set point
- Peak Time = time to the first peak

- Settling Time = time to when the PV first enters and then remains within a band whose width is computed as a percentage of the total change in PV (or DPV).

The 5% band used to determine settling time in the plot above was chosen arbitrarily. As we will see below, other percentages are equally valid depending on the situation.

From the plot, we see that the set point is stepped at time t = 30 min. The time related criteria are then computed by reading off the time axis as:

Rise Time = (43 – 30) = 13 min

Peak Time = (51 – 30) = 21 min

Settling Time = (100 – 30) = 70 min for a ±5% of DPV band

When there is no Overshoot

We should recognize that the peak and time criteria are not independent:

- a process with a large decay ratio will likely have a long settling time.
- a process with a long rise time will likely have a long peak time.

And in situations where we seek moderate tuning with no overshoot in our response plots, there is no peak overshoot ratio, decay ratio or peak time to compute. Even rise time, with its asymptotic approach to the new steady state, is a measure of questionable value.

In such cases, settling time, or the time to enter and remain within a band of width we choose, still remains a useful measure.

The plot below shows the identical process as that in the previous plots. The only difference is that in this case, the controller is tuned for a moderate response.

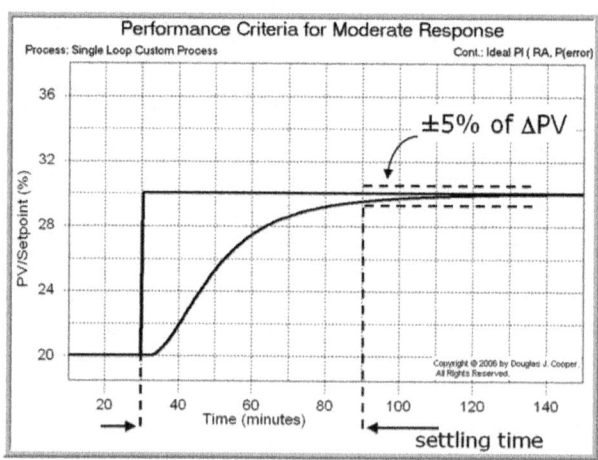

We compute for this plot:

Settling Time = (90 – 30) = 60 min for a ±5% of DPV band

Isn't it interesting that a more moderately tuned controller has a faster settling time then the more active or aggressive response shown previously where settling time was 70 minutes? (In truth, this is not a surprising result because for the previous plots, the controller was deliberately mistuned to provide the sharp peaks we needed to clearly illustrate the performance definitions.)

Settling Band is Application Specific

Measurement noise, or random error in the PV signal, can be one reason to modify the width of the settling band. As shown below, for the identical process and tuning as used in the above moderate response plot, but with significant noise added to the PV, the settling band must to be widened considerably to provide a meaningful measure.

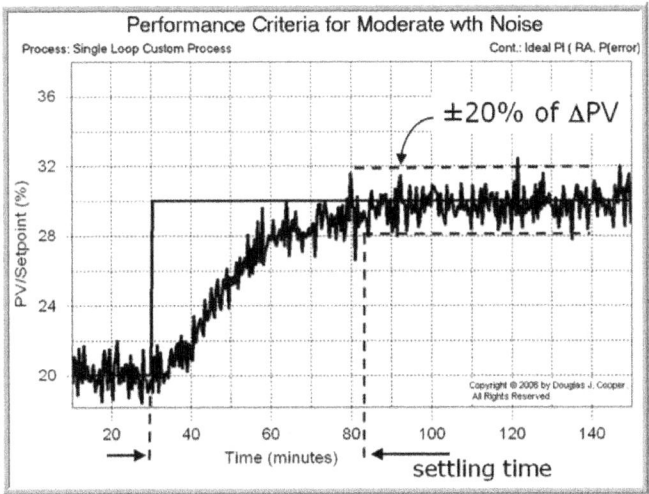

Widening the band in this case is not a bad thing. It is simply necessary. And since we will (presumably) use the same settling band criteria when exploring alternatives for this loop, then it remains a useful tool for comparing performance.

Chapter 7

THE EVALUATION AND CONTROL OF ORGANIZATIONAL STRATEGY

The basic premise of strategic management is that the chosen strategy will achieve the organization's mission and objectives.

A firm's successive strategies are greatly affected by its past history and often take shape through experimentation and ad hoc refinement of current plans, a process *James Quinn* has termed "**logical incrementalism**". Therefore, the reexamination of past assumptions, the comparison of actual results with earlier hypotheses have become common features of strategic management.

This chapter describes the nature of control, strategic controls and explains a how to set them up. It them explains key operational control systems necessary to support strategic control.

Nature of Control

Management control refers to the process by which an organization influences its subunits and members to behave in ways that lead to the attainment of organizational objectives.

What is Control?

Management control is a systematic effort to set performance standards with planning objectives, to design information feedback systems, to compare actual performance with these predetermined standards, to determine whether there are any deviations and to measure their significance, and to take any action required to assure that all corporate resources are being used in the most effective and efficient way possible in achieving corporate objectives.

Types of Control

Management can implement controls before an activity commences, while the activity is going on, or after the activity has been completed. The three respective types of control based on timing are feedforward, concurrent, and feedback.

Feedforward Control

Feedforward control focuses on the regulation of inputs (human, material, and financial resources that flow into the organization) to ensure that they meet the standards necessary for the transformation process.

Feedforward controls are desirable because they allow management to prevent problems rather than having to cure them later. Unfortunately, these control require timely and accurate information that is often difficult to develop. Feedforward control also is sometimes called **preliminary control**, **precontrol**, **preventive control**, or **steering control**.

However, some authors use term "*steering control*" as separate types of control. This types of controls are designed to detect deviation some standard or goal to allow correction to be made before a particular sequence of actions is completed.

Concurrent Control

Concurrent control takes place while an activity is in progress. It involves the regulation of ongoing activities that are part of transformation process to ensure that they conform to organizational standards. Concurrent control is designed to ensure that employee work activities produce the correct results.

Since concurrent control involves regulating ongoing tasks, it requires a through understanding of the specific tasks involved and their relationship to the desired and product.

Concurrent control sometimes is called **screening** or **yes-no control**, because it often involves checkpoints at which determinations are made about whether to continue progress, take corrective action, or stop work altogether on products or services.

Feedback Control

This type of control focuses on the outputs of the organization after transformation is complete. Sometimes called **postaction** or **output control**, fulfils a number of important functions. For one thing, it often is used when feedforward and concurrent controls are not feasible or are to costly.

Sometimes, feedback is the only viable type of control available. Moreover, feedback has two advantages over feedforward and concurrent control. *First*, feedback provides managers with meaningful information on how effective its planning effort was. If feedback indicates little variance between standard and actual performance, this is evidence that planning was generally on target.

If the deviation is great, a manager can use this information when formulating new plans to make them more effective. *Second,* feedback control can enhance employees motivation.

The major drawback of this type of control is that, the time the manager has the information and if there is significant problem the damage is already done. But for many activities, feedback control fulfils a number important functions.

Multiple Controls

Feedforward, concurrent, and feedback control methods are not mutually exclusive. Rather, they usually are combined into an multiple control systems. Managers design control systems to define standards of performance and acquire information feedback at strategic control points.

Strategic control points are those activities that are especially important for achieving strategic objectives. When organizations do not have multiple control systems that focus on strategic control points, they often can experience difficulties that cause managers to reevaluate their control processes.

Managerial Approaches to Implementing Controls

Regardless of whether the organization focuses control on inputs, production, or outputs, another choice must be made between different approaches tor control. There are three control approaches regarding the mechanisms managers will use to implement controls: *market control, bureaucratic control, and clan control.*

Market Control

Market control involves the use of price competition to evaluate output. Managers compare profits and prices to determine the efficiency of their organization. In order to use market control, there must be a reasonable level of competition in the goods or service area and it must be possible to specify requirements clearly.

Market control is non appropriate in controlling functional departments, unless the price for services is set through competition and its representative of the true value of provided services.

Bureaucratic Control

Bureaucratic control is the use of rules, policies, hierarchy of authority, written documentation, reward systems, and other formal mechanisms to influence employee behavior and assess performance. Bureaucratic control can be used when behavior can be controlled with market or price mechanisms.

Clan Control

Clan control represents cultural values almost the opposite of bureaucratic control. Clan control relies on values, beliefs, corporate culture, shared norms, and informal relationships to regulate employee behaviors and facilitate the reaching of organizational goals.

Organization that use clan control require trust among their employees. Given minimal direction and standards, employees are assumed to perform well - indeed, they participate in setting standards and designing the control systems.

The Primary Types of Organizational Control

There are three primary types of organizational control: *strategic control, management control, and operational control*.

Strategic control, the process of evaluating strategy, is practiced both after the strategy is formulated and after it is implemented.

Management control focuses on the accomplishment of the objectives of the various substrategies comprising the master strategy and the accomplishment of the objectives of the intermediate plans (for example, *"are quality control objectives being met?"*).

Operational control is concerned individual and group performance as compared with the individual and group role prescriptions required by organizational plans (for example, *"are individual sales quotes being met?"*).

Each of these types of control is not a separate and distinct entity and, in fact, may be indistinguishable from others. Moreover, similar measurement techniques may be used for each type of control.

Strategic Control

Strategic control is concerned with tracking the strategy as it is being implemented, detecting any problems areas or potential problem areas, and making any necessary adjustments.

Newman and Logan use the term "**steering control**" to highlight some important characteristics of strategic control Ordinarily, a significant time span occurs between initial implementation of a strategy and achievement of its intended results. During that time, numerous projects are undertaken, investments are made, and actions are undertaken to implement the new strategy.

Also the environmental situation and the firm's internal situation are developing and evolving. Strategic controls are necessary to steer the firm through these events. They must provide some means of correcting the directions on the basis of intermediate performance and new information.

The Importance of Strategic Control

Henry Mintzberg, one of the foremost theorists in the area of strategic management, tells us that no matter how well the organization plans its strategy, a different strategy may emerge.

Starting with the intended or planned strategies, he related the five types of strategies in the following manner:

1. Intended strategies that get realized; these may be called **deliberate** strategies.

2. Intended strategies that do get realized; these may be called **unrealized** strategies.

3. Realized strategies that were never intended; these may be called **emergent** strategies.

Recognizing the number of different ways that intended and realized strategies may differ underscores the importance of evaluation and control systems so that the firm can monitor its performance and take corrective action if the actual performance differs from the intended strategies and planned results.

Management Control

Where management control is imposed, it functions within the framework established by the strategy. Normally these objectives (standards) are established for major subsystems within the organization, such as *SBUs*, projects, products, functions, and responsibility centers.

Typical management control measures include *ROI*, residual income, cost, product quality, and so on. These control measures are essentially summations of operational control measures. Corrective action may involve very minor or very major changes in the strategy.

Operating Control

Operational control systems are designed to ensure that day-to-day actions are consistent with established plans and objectives. It focuses on events in a recent period. Operational control systems are derived from the requirements of the management control system.

Corrective action is taken where performance does not meet standards. This action may involve training, motivation, leadership, discipline, or termination.

Differences between Strategic and Operational Control

The differences between strategic and operational control are highlighted by reference to a general definition of management control: "**Management control is the set of measurement, analysis, and action decisions required for the timely management of the continuing operation of a process**".

Measurement:

- **Strategic control requires data from more sources**. The typical operational control problem uses data from very few sources.

- **Strategic control requires more data from external sources**. Strategic decisions are normally taken with regard to the external environment as opposed to internal operating factors.

- **Strategic control are oriented to the future**. This is in contrast to operational control decisions in which control data give rise to immediate decisions that have immediate impacts.

- **Strategic control is more concerned with measuring the accuracy of the decision premise**. Operating decisions tend to be concerned with the quantitative value of certain outcomes.

- **Strategic control standards are based on external factors**. Measurement standards for operating problems can be established fairly by past performance on similar products or by similar operations currently being performed.

- **Strategic control relies on variable reporting interval**. The typical operating measurement is concerned with operations over some period of time: pieces per week, profit per quarter, and the like.

Analysis:

- **Strategic control models are less precise**. This is in contrast to operational control models, which are generally very precise in the narrow domain they apply.

- **Strategic control models are less formal**. The models that govern the considerations in a strategic control problem are much more intuitive, therefore, less formal.

- **The principal variables in a strategic control model are structural**. In strategic control, the whole structure of the problem, as represented by the model, is likely to vary, not just the values of the parameters.

- **The key need in analysis for strategic control is model flexibility**. This is in contrast to operating control, for which efficient quantitative computation is usually most desirable.

- **The key activity in management control analysis is alternative generation**. This is different from the operational control problem, in which in many cases all control alternatives have been specified in advance. The key analysis step in operations is to discover exactly what happened.

- **The key skill required for management control analysis is creativity**. In operational control, by contrast, the formal review of outcomes to discover causes means that they skill required is the ability to do technical, even statistical, analysis of the data received.

Action:

- **The relationship between action and outcome is weaker in strategic control**. This is not surprising, as the most desirable area for control in strategic problems -the environment -is the least subject to direct action.

- **The key action variables in strategic control are organizational.** In the operational control problem, technical factors such as labor levels, production levels, choice of materials, and the like are the predominant control levels.

- **Alternative actions in strategic control are less easy to choose in advance.** In strategic control problem, it is possible to choose all possible action responses to received data in advance. In an operational control problem, the few responses possible can usually all be worked out before any operating data received.

- **The worst failing in strategic control is omitting a worthwhile action.** In operating control, the most typical sins are those of omissions (*e.g.*, complaints about too many people employed, too many defects, and too much inventory). In the strategic control problem, sins of omission are much more serious (*e.g.*, not moving into a business opportunity when it presents itself, not undertaking a particular social program, not applying resources to meet that challenges in the best fashion).

- **The time for strategic control is longer.** The period in which control has an impact is longer for strategic problems that for operating problems.

- **The timing of strategic control is events oriented.** By contrast, operating decisions tend to be made on a periodic basis, and they are usually measured accordingly.

- **Strategic control has little repetition.** Not even the structure is the same as past problems of a like kind, much less the technical details. Operating problems, by way of contrast, tend to repeat their structure.

Implications for Information Systems

- **Strategic control requires a greater variety of data types.** Operating control problems typically have a smaller variety of data.

- **The total volume of data required for strategic control is smaller.** On the other hand, perhaps thousands of pieces of data of each type are required for some of operating problems (*e.g.*, the payroll processing of even a small organization).

- **Strategic control data are more aggregated.** Operating data are used at the most detailed at transaction level.

- **Strategic control data are less accurate.** Operating data generally need to be as accurate as possible.

- **The most important strategic control information is structural.** Unlike the operational control are, the values of the technical variables are only of secondary importance.

- **The receipt of data for strategic control is more sporadic.** Data for strategic problems are received sporadically as events take place.

- **Strategic control data are less processable by computer**. The strategic control that arise in the environment rather than within the organization are generally not so easily available. For the most part, such data need not be computerized. It does imply that any computerization of strategic control tools must consider the important step of capturing necessary in machine-readable form.

- **The key decision in information for strategic control is what data to save**. The principal problem in operating control information systems design is the technological problem of efficiently capturing and retrieving data.

Implications for Controlling Formal Plans

- **Contingency plans are less possible in strategic control**. The whole idea of contingency plans is much more difficult in the strategic arena. It is more difficult to generate all possible actions ahead of time in a strategic problem, because the alternatives are too numerous and too complex.

- **Triggering contingency planning is more important in strategic control**. Because of this difficulty in making contingency plans, triggering an examination of alternatives when things do not go according to plan becomes much more important.

- **Preprogrammed variance analysis is less possible in strategic control**. For an operational control model might be possible that the computer performs all possible variance analyses (in the accounting sense). For strategic control it is both difficult technically and impossible practically.

- **A variance inquiry system is more necessary in strategic control**. It seems important to have an inquiry system linked to the formal planning model with which combinations of deviations from plans can be explored by the human operator.

- **A variance inquiry language is more necessary in strategic control**. Some sort of language in which the human can do variance inquiries is highly desirable in the area of strategic control.

- **An augmented formal planning system in more necessary in strategic control**. A formal planning system should be augmented with the variance inquiry language described. This would permit the same system that was used to generate the plan to be used in controlling that plan, leading to both ease of additional analysis as well, as to consistency with the plan being controlling.

Approaches to the Evaluation Organizational Effectiveness

An organization's effectiveness is in major part a measure of the effectiveness of its master strategy. Selection of the appropriate basis for assessing organizational effectiveness presents a challenging problem for managers and researchers.

There are no generally accepted conceptualizations prescribing the best criteria. Different organizational situations - pertaining to the performance of the organization's structure, the performance of the organization's human resources, and the impact of the organization's activities -require different criteria.

J. Barton Cunningham, after reviewing the relevant literature, concluded that *seven major ways* of evaluating organizational effectiveness existed: *rational goal model, systems resource model, managerial process model, organizational development model, the bargaining model*.

The Rational Goal Model

The rational goal approach focuses on the organization's ability to achieve its goals. An organization's goals are identified by establishing the general goal, discovering means or objectives for its accomplishment, and defining a set of activities for each objectives.

The organization is evaluated by comparing the activities accomplished with those planned for. These criteria are determined by various factors.

The Systems Resource Model

The systems resource model analyzes the decision-makers's capability to efficiently distribute resources among various subsystem's needs. The systems resources model defines the organization as a network of interrelated subsystems.

These subsystems needs may be classified as:

- *bargaining position* -ability of the organization to exploit its environment in acquisition of scarce and valued resources;
- ability of the systems' decision-makers to perceive, and correctly interpret, the real properties of the external environment;
- ability of the system to produce a certain specified output;
- maintenance of internal day-to-day activities;
- ability of the organization to co-ordinate relationships among the various subsystems;
- ability of the organization to respond to feedback regarding its effectiveness in the environment.
- ability of the organization to evaluate the effect of its decisions;
- ability of the organization' system to accomplish its goals.

The Bargaining Model

Each organizational problem requires a specific allocation of resources. The bargaining model presumes that an organization is a cooperative, sometimes competitive, resource distributing system.

Decisions, problems and goals are more useful when shared by a greater number of people. Each decision-maker bargains with other groups for scarce resources which are vital in solving problems and meeting goals.

The overall outcome is a function of the particular strategies selected by the various decision-makers in their bargaining relationships. This model measures the ability of decision-makers to obtain and use resources for responding to problems important to them.

Each of the subsystems' needs should be evaluated from two focal points: efficiency and stress. *Efficiency* is an indication of the organization's ability to use its resources in responding to the most subsystems' needs. Stress is the tension produced by the system in fulfilling or not fulfilling its needs.

The Managerial Process Model

The managerial process model assesses the capability an productivity of various managerial processes -decision making, planning, budgeting, and the like -for performing goals.

The managerial process model is based on the intuitive concept of substantial rationality, which interrelates the drives, impulses, wishes, feelings, needs, and values of the individuals to the functional goals of the organization.

THE ORGANIZATIONAL DEVELOPMENT MODEL

This model appraises the organization's ability to work as a team and to fit the needs of its members. The model focuses on developing practices to foster:

1. supervisory behavior manifesting interest and concern for workers;
2. team spirit, group loyalty, and teamwork among workers and between workers and management;
3. confidence, trust and communication among workers and between workers and management;
4. more freedom to set their own objectives.

The model's procedure attempts to answer four main questions:

1. Where are we?;
2. Where do we want to go?;
3. How will we get there?;
4. How will we know when we do get there?

These questions can be divided into four areas: question one is concerned with diagnosis, question two with the setting of goals and plans, question three with the implementation of goals, and question four with evaluation.

This model is concerned with changing beliefs, attitudes, values, and organizational structures so that individuals can be better adopt to new technologies and challenges. It is a process of management by objectives in contrast to management by control.

The Structural Functional Model

The structural functional approach tests the durability and flexibility of the organization's structure for responding to a diversity of situations and events.

According to this model, all systems need maintenance and continuity. The following aspects define this:

- security of the organization as whole in relation to the social forces in its environment (this relates to ability to forestall threatened aggressions or deleterious consequences from the actions of others);
- stability of lines of authority and communication (this refers to the continued capacity of leadership to control and have access to individuals in the system);
- stability of informal relations within the organization;
- continuity of policy making (this refers to the ability to reexamine policy an a continuing basis);
- homogeneity of outlook (this refers the ability to effectively orient members to organization norms and beliefs).

The Functional Model

In the functional approach an organization's effectiveness is determined by the social consequences of its activities.

The crucial question to be answered is:**how well do the organization's activities serve the needs of its client groups?**

The appraisal of an organization's effectiveness should consider whether these activities are function or dysfunctions in fulling the organization's goals.

These seven models have their strengths and shortcomings depending upon the organizational situation being evaluated. The choice of evaluation approach usually hinges on the organizational situation that needs to be addressed.

Anthony's Perspective

Until 1965, and not uncommonly even today, conventional wisdom held that planning and control in organizations should be separated: "**...** **control must reflect plans; and planning must precede control**". In 1965, *Robert Anthony* of the Harvard Business School put forward a novel framework for the analysis of planning and control systems.

Anthony's basic thesis is that planning and control are so closely interlinked in organizations as to make their separation meaningless and undesirable. He suggests, it makes much more conceptual and practical sense to link together similar and interwind planning and control activities into systems of homogeneous characteristics.

Instead of two categories of planning and control (a practice still supported by certain authorities), Anthony suggests that organizational planning and control be segmented into three categories:

1. strategic planning
2. management control
3. task (operational) control

It is important to note that Anthony's terminology, is somewhat misleading. When Anthony says *"strategic planning"* he means *"strategic planning and control."* Similarly *"management control"* embraces both planning and control activities.

Strategic Planning

According to Anthony:

"Strategic planning is the process of deciding on the goals of the organization and the strategies for attaining these goals."

Strategies are guidelines for deciding the appropriate actions for attaining the organization's goals. The essential difference between strategic planning and management control is that the strategic planning process is unsystematic.

Strategic control occurs in three ways. *First,* strategic planning is itself a form of control. *Second,* strategic plans are converted into reality not only by their influence on the management control activity but also by the key decisions regarding allocation of resources.

Third, while capital budgeting systems can respond to requests for resources that are consistent with the accepted strategic plan, the period between formal, comprehensive strategic planning exercises can give rise to unanticipated changes in the environment or unexpected internal crises.

Anthony views management planning and control as the processes by which:

(**1**) organizational objectives are achieved and

(**2**) the use of resources is made effective and efficient.

"Management control is the process by which managers influence other members of the organization to implement the organization's strategies."

Management control decisions are made within the guidance established by strategic planning. Management control is a systematic process. It is done by managers at all levels; it is done on regular basis; it involves the whole organization; and it involves a large amount of personal interaction and relatively less judgment.

There are two somewhat different types of management control activities:

(**1**) *the management control of operating activities,* and

(**2**) *the control of operational projects.*

Process for operating activities has four phases: *programming, budget preparation, execution, and evaluation.*

Programming is the process of deciding on the major programs that the organization will undertake to implement its strategies and the approximate amount of resources that will be devoted to each.

Budget preparation. An operating budget is the organization's financial plan for a specific period, usually one year.

Execution and evaluation. During the year managers execute the program or part of a program for which they are responsible. Reports on responsibility centers show both budgeted and actual information. They are used as a basis for control. The process of evaluation is a comparison of actual amounts with the amounts that should be expected of actual circumstances.

A projects is a set of activities intended to accomplish a specified end result of sufficient importance to be of interest to management (for example: construction projects, research/development projects, and motion picture productions).

In a project, and in each of its components, the focus is on three aspects:

(1) its scope (that is, the specifications for the end product),

(2) its schedule (that is, the time required), and

(3) its cost.

In actual operations, project managers engage in both planning activities and control activities. They control when they act to improve effectiveness and efficiency.

Anthony views this third category of organizational planning and control as

(1) focusing on specific, discrete tasks and

(2) the process of ensuring that those tasks are done effectively and efficiently.

"Task control is the process of ensuring that specific tasks are carried out effectively and efficiently."

As the definition suggests, the focus of operational control is on individual tasks or transaction: scheduling and controlling individual jobs through a shop, as contrasted with measuring the performance of the shop as a whole; procuring specific items for inventory, as contrasted with management of inventory as whole: and so on.

Task control is distinguished from management control in the following ways:

- The management control system is basically of similar throughout the organization. Each type task requires a different task control system.

- In management control, managers interact with other managers; in task control either humans are not involved at all, or the interaction is between a manager and a nonmanager.

- In management control the focus is on organizational units called responsibility centers; in task control the focus is on specific tasks.

- Management control relates to activities that are not specified; task control relates to specified tasks.

- In management control the focus is equally on planning and on execution; in task control it is primarily on execution.

An essential characteristic of the process is that the "*standard*" against which actual performance is measured is consistent with the organization's strategies. *Exhibit* outlines differences among the three types of processes with respect to the nature of the problems that typically are addressed in each process and the types of decisions that are relevant for these problems.

As another way of explaining the differences among the three processes, *Exhibit* gives some examples of activities associated with each.

Most commentators would agree with the definition of strategic control offered by *Schendel and Hofer*:

"**Strategic control focuses on the dual questions of whether:**

(1) the strategy is being implemented as planned; and

(2) the results produced by the strategy are those intended."

This definition refers to the traditional review and feedback stages which constitutes the last step in the strategic management process. Normative models of the strategic management process have depicted it as including there primary stages: strategy formulation, strategy implementation, and strategy evaluation (control).

Strategy evaluations concerned primarily with traditional controls processes which involves the review and feedback of performance to determine if plans, strategies, and objectives are being achieved, with the resulting information being used to solve problems or take corrective actions.

Recent conceptual contributors to the strategic control literature have argued for anticipatory feedforward controls, that recognize a rapidly changing and uncertain external environment.

Schreyogg and Steinmann (1987) have made a preliminary effort, in developing new system to operate on a continuous basis, checking and critically evaluating assumptions, strategies and results. They refer to strategic control as "**the critical evaluation of plans, activities, and results, thereby providing information for the future action**".

Schreyogg and Steinmann based on the shortcomings of feedback-control. Two central characteristics if this feedback control is highly questionable for control purposes in strategic management: (**a**) *feedback control is post-action control and* (**b**) *standards are taken for granted.*

Schreyogg and Steinmann proposed an alternative to the classical feedback model of control: a 3-step model of strategic control which includes **premise control, implementation control**, and **strategic surveillance**. *Pearce and Robin-*

son extended this model and added a component "**special alert control**" to deal specifically with low probability, high impact threatening events.

The nature of these four strategic controls is summarized in *Figure*. Time (*t*) marks the point where strategy formulation starts. Premise control is established at the point in time of initial premising (*t*). From here on promise control accompanies all further selective steps of premising in planning and implementing the strategy. The strategic surveillance of emerging events parallels the strategic management process and runs continuously from time (*t*) through (*t*). When strategy implementation begins (*t*), the third control device,*implementation control* is put into action and run through the end of the planning cycle (*t*). Special alert controls are conducted over the entire planning cycle.

Planning premises/assumptions are established early on in the strategic planning process and act as a basis for formulating strategies.

"Premise control has been designed to check systematically and continuously whether or not the premises set during the planning and implementation process are still valid.

It involves the checking of environmental conditions. Premises are primarily concerned with two types of factors:

- **Environmental factors** (for example, inflation, technology, interest rates, regulation, and demographic/social changes).
- **Industry factors** (for example, competitors, suppliers, substitutes, and barriers to entry).

All premises may not require the same amount of control. Therefore, managers must select those premises and variables that (**a**)are likely to change and (**b**) would a major impact on the company and its strategy if the did.

Management Control

Where management control is imposed, it functions within the framework established by the strategy. Normally these objectives (standards) are established for major subsystems within the organization, such as *SBUs*, projects, products, functions, and responsibility centers.

Typical management control measures include *ROI*, residual income, cost, product quality, and so on. These control measures are essentially summations of operational control measures. Corrective action may involve very minor or very major changes in the strategy.

Operational control systems are designed to ensure that day-to-day actions are consistent with established plans and objectives. It focuses on events in a recent period. Operational control systems are derived from the requirements of the management control system.

Corrective action is taken where performance does not meet standards. This action may involve training, motivation, leadership, discipline, or termination.

The differences between strategic and operational control are highlighted by reference to a general definition of management control: "**Management control is the set of measurement, analysis, and action decisions required for the timely management of the continuing operation of a process**". This section discusses in the terms presented.

- **Strategic control requires data from more sources.** The typical operational control problem uses data from very few sources.

- **Strategic control requires more data from external sources.** Strategic decisions are normally taken with regard to the external environment as opposed to internal operating factors.

- **Strategic control are oriented to the future.** This is in contrast to operational control decisions in which control data give rise to immediate decisions that have immediate impacts.

- **Strategic control is more concerned with measuring the accuracy of the decision premise.** Operating decisions tend to be concerned with the quantitative value of certain outcomes.

- **Strategic control standards are based on external factors.** Measurement standards for operating problems can be established fairly by past performance on similar products or by similar operations currently being performed.

- **Strategic control relies on variable reporting interval.** The typical operating measurement is concerned with operations over some period of time: pieces per week, profit per quarter, and the like.

- **Strategic control models are less precise.** This is in contrast to operational control models, which are generally very precise in the narrow domain they apply.

- **Strategic control models are less formal.** The models that govern the considerations in a strategic control problem are much more intuitive, therefore, less formal.

- **The principal variables in a strategic control model are structural.** In strategic control, the whole structure of the problem, as represented by the model, is likely to vary, not just the values of the parameters.

- **The key need in analysis for strategic control is model flexibility.** This is in contrast to operating control, for which efficient quantitative computation is usually most desirable.

- **The key activity in management control analysis is alternative generation.** This is different from the operational control problem, in which in many cases all control alternatives have been specified in advance. The key analysis step in operations is to discover exactly what happened.

- **The key skill required for management control analysis is creativity.** In operational control, by contrast, the formal review of outcomes to discover causes means that they skill required is the ability to do technical, even statistical, analysis of the data received.

- **The relationship between action and outcome is weaker in strategic control.** This is not surprising, as the most desirable area for control in strategic problems -the environment -is the least subject to direct action.

- **The key action variables in strategic control are organizational.** In the operational control problem, technical factors such as labor levels, production levels, choice of materials, and the like are the predominant control levels.

- **Alternative actions in strategic control are less easy to choose in advance.** In strategic control problem, it is possible to choose all possible action responses to received data in advance. In an operational control problem, the few responses possible can usually all be worked out before any operating data received.

- **The worst failing in strategic control is omitting a worthwhile action.** In operating control, the most typical sins are those of omissions (*e.g.,* complaints about too many people employed, too many defects, and too much inventory). In the strategic control problem, sins of omission are much more serious (*e.g.,* not moving into a business opportunity when it presents itself, not undertaking a particular social program, not applying resources to meet that challenges in the best fashion).

- **The time for strategic control is longer.** The period in which control has an impact is longer for strategic problems that for operating problems.

- **The timing of strategic control is events oriented.** By contrast, operating decisions tend to be made on a periodic basis, and they are usually measured accordingly.

- **Strategic control has little repetition.** Not even the structure is the same as past problems of a like kind, much less the technical details. Operating problems, by way of contrast, tend to repeat their structure.

- **Strategic control requires a greater variety of data types.** Operating control problems typically have a smaller variety of data.

- **The total volume of data required for strategic control is smaller.** On the other hand, perhaps thousands of pieces of data of each type are required for some of operating problems (*e.g.,* the payroll processing of even a small organization).

- **Strategic control data are more aggregated.** Operating data are used at the most detailed at transaction level.

- **Strategic control data are less accurate.** Operating data generally need to be as accurate as possible.

- **The most important strategic control information is structural.** Unlike the operational control are, the values of the technical variables are only of secondary importance.

- **The receipt of data for strategic control is more sporadic**. Data for strategic problems are received sporadically as events take place.

- **Strategic control data are less processable by computer**. The strategic control that arise in the environment rather than within the organization are generally not so easily available. For the most part, such data need not be computerized. It does imply that any computerization of strategic control tools must consider the important step of capturing necessary in machine-readable form.

- **The key decision in information for strategic control is what data to save**. The principal problem in operating control information systems design is the technological problem of efficiently capturing and retrieving data.

- **Contingency plans are less possible in strategic control**. The whole idea of contingency plans is much more difficult in the strategic arena. It is more difficult to generate all possible actions ahead of time in a strategic problem, because the alternatives are too numerous and too complex.

- **Triggering contingency planning is more important in strategic control**. Because of this difficulty in making contingency plans, triggering an examination of alternatives when things do not go according to plan becomes much more important.

- **Preprogrammed variance analysis is less possible in strategic control**. For an operational control model might be possible that the computer performs all possible variance analyses (in the accounting sense). For strategic control it is both difficult technically and impossible practically.

- **A variance inquiry system is more necessary in strategic control**. It seems important to have an inquiry system linked to the formal planning model with which combinations of deviations from plans can be explored by the human operator.

- **A variance inquiry language is more necessary in strategic control**. Some sort of language in which the human can do variance inquiries is highly desirable in the area of strategic control.

- **An augmented formal planning system in more necessary in strategic control**. A formal planning system should be augmented with the variance inquiry language described. This would permit the same system that was used to generate the plan to be used in controlling that plan, leading to both ease of additional analysis as well, as to consistency with the plan being controlling.

An organization's effectiveness is in major part a measure of the effectiveness of its master strategy. Selection of the appropriate basis for assessing organizational effectiveness presents a challenging problem for managers and researchers.

There are no generally accepted conceptualizations prescribing the best criteria. Different organizational situations - pertaining to the performance of the organization's structure, the performance of the organization's human resources, and the impact of the organization's activities -require different criteria.

J. Barton Cunningham, after reviewing the relevant literature, concluded that *seven major ways* of evaluating organizational effectiveness existed: *rational goal model, systems resource model, managerial process model, organizational development model, the bargaining model.*

The rational goal approach focuses on the organization's ability to achieve its goals. An organization's goals are identified by establishing the general goal, discovering means or objectives for its accomplishment, and defining a set of activities for each objectives.

The organization is evaluated by comparing the activities accomplished with those planned for. These criteria are determined by various factors.

The systems resource model analyzes the decision-makers's capability to efficiently distribute resources among various subsystem's needs. The systems resources model defines the organization as a network of interrelated subsystems.

These subsystems needs may be classified as:

- *bargaining position* -ability of the organization to exploit its environment in acquisition of scarce and valued resources;
- ability of the systems' decision-makers to perceive, and correctly interpret, the real properties of the external environment;
- ability of the system to produce a certain specified output;
- maintenance of internal day-to-day activities;
- ability of the organization to co-ordinate relationships among the various subsystems;
- ability of the organization to respond to feedback regarding its effectiveness in the environment.
- ability of the organization to evaluate the effect of its decisions;
- ability of the organization' system to accomplish its goals.

Each organizational problem requires a specific allocation of resources. The bargaining model presumes that an organization is a cooperative, sometimes competitive, resource distributing system.

Decisions, problems and goals are more useful when shared by a greater number of people. Each decision-maker bargains with other groups for scarce resources which are vital in solving problems and meeting goals.

The overall outcome is a function of the particular strategies selected by the various decision-makers in their bargaining relationships. This model measures the ability of decision-makers to obtain and use resources for responding to problems important to them.

Each of the subsystems' needs should be evaluated from two focal points: efficiency and stress. *Efficiency* is an indication of the organization's ability to use its resources in responding to the most subsystems' needs. Stress is the tension produced by the system in fulfilling or not fulfilling its needs.

The managerial process model assesses the capability an productivity of various managerial processes -decision making, planning, budgeting, and the like -for performing goals.

The managerial process model is based on the intuitive concept of substantial rationality, which interrelates the drives, impulses, wishes, feelings, needs, and values of the individuals to the functional goals of the organization.

This model appraises the organization's ability to work as a team and to fit the needs of its members. The model focuses on developing practices to foster:

1. supervisory behavior manifesting interest and concern for workers;

2. team spirit, group loyalty, and teamwork among workers and between workers and management;

3. confidence, trust and communication among workers and between workers and management;

4. more freedom to set their own objectives.

The model's procedure attempts to answer four main questions:

1. Where are we?;

2. Where do we want to go?;

3. How will we get there?;

4. How will we know when we do get there?

These questions can be divided into four areas: question one is concerned with diagnosis, question two with the setting of goals and plans, question three with the implementation of goals, and question four with evaluation.

This model is concerned with changing beliefs, attitudes, values, and organizational structures so that individuals can be better adopt to new technologies and challenges. It is a process of management by objectives in contrast to management by control.

The structural functional approach tests the durability and flexibility of the organization's structure for responding to a diversity of situations and events.

According to this model, all systems need maintenance and continuity. The following aspects define this:

- security of the organization as whole in relation to the social forces in its environment (this relates to ability to forestall threatened aggressions or deleterious consequences from the actions of others);

- stability of lines of authority and communication (this refers to the continued capacity of leadership to control and have access to individuals in the system);

- stability of informal relations within the organization;

- continuity of policy making (this refers to the ability to reexamine policy an a continuing basis);

- homogeneity of outlook (this refers the ability to effectively orient members to organization norms and beliefs).

In the functional approach an organization's effectiveness is determined by the social consequences of its activities.

The crucial question to be answered is: **how well do the organization's activities serve the needs of its client groups?**

The appraisal of an organization's effectiveness should consider whether these activities are function or dysfunctions in fulling the organization's goals.

These seven models have their strengths and shortcomings depending upon the organizational situation being evaluated. The choice of evaluation approach usually hinges on the organizational situation that needs to be addressed.

Until 1965, and not uncommonly even today, conventional wisdom held that planning and control in organizations should be separated: "**... control must reflect plans; and planning must precede control**". In 1965, *Robert Anthony* of the Harvard Business School put forward a novel framework for the analysis of planning and control systems.

Anthony's basic thesis is that planning and control are so closely interlinked in organizations as to make their separation meaningless and undesirable. He suggests, it makes much more conceptual and practical sense to link together similar and interwind planning and control activities into systems of homogeneous characteristics.

Instead of two categories of planning and control (a practice still supported by certain authorities), Anthony suggests that organizational planning and control be segmented into three categories:

1. strategic planning
2. management control
3. task (operational) control

It is important to note that Anthony's terminology, is somewhat misleading. When Anthony says "*strategic planning*" he means "*strategic planning and control.*" Similarly "*management control*" embraces both planning and control activities.

According to Anthony:

"Strategic planning is the process of deciding on the goals of the organization and the strategies for attaining these goals."

Strategies are guidelines for deciding the appropriate actions for attaining the organization's goals. The essential difference between strategic planning and management control is that the strategic planning process is unsystematic.

Strategic control occurs in three ways. *First*, strategic planning is itself a form of control. *Second*, strategic plans are converted into reality not only by their influence on the management control activity but also by the key decisions regarding allocation of resources.

Third, while capital budgeting systems can respond to requests for resources that are consistent with the accepted strategic plan, the period between formal, comprehensive strategic planning exercises can give rise to unanticipated changes in the environment or unexpected internal crises.

Anthony views management planning and control as the processes by which :

(1) organizational objectives are achieved and

(2) the use of resources is made effective and efficient.

"Management control is the process by which managers influence other members of the organization to implement the organization's strategies."

Management control decisions are made within the guidance established by strategic planning. Management control is a systematic process. It is done by managers at all levels; it is done on regular basis; it involves the whole organization; and it involves a large amount of personal interaction and relatively less judgment.

There are two somewhat different types of management control activities:

(1) *the management control of operating activities*, and

(2) *the control of operational projects*.

The Management Control

Process for operating activities has four phases: *programming, budget preparation, execution, and evaluation.*

Programming is the process of deciding on the major programs that the organization will undertake to implement its strategies and the approximate amount of resources that will be devoted to each.

Budget preparation. An operating budget is the organization's financial plan for a specific period, usually one year.

Execution and evaluation. During the year managers execute the program or part of a program for which they are responsible. Reports on responsibility centers show both budgeted and actual information. They are used as a basis for control. The process of evaluation is a comparison of actual amounts with the amounts that should be expected of actual circumstances.

MANAGEMENT CONTROL OF PROJECTS

A projects is a set of activities intended to accomplish a specified end result of sufficient importance to be of interest to management (for example: construction projects, research/development projects, and motion picture productions).

In a project, and in each of its components, the focus is on three aspects:

(1) its scope (that is, the specifications for the end product),

(2) its schedule (that is, the time required), and (

(3) its cost.

In actual operations, project managers engage in both planning activities and control activities. They control when they act to improve effectiveness and efficiency.

Task Control

Anthony views this third category of organizational planning and control as:

(1) focusing on specific, discrete tasks and

(2) the process of ensuring that those tasks are done effectively and efficiently.

"Task control is the process of ensuring that specific tasks are carried out effectively and efficiently."

As the definition suggests, the focus of operational control is on individual tasks or transaction: scheduling and controlling individual jobs through a shop, as contrasted with measuring the performance of the shop as a whole; procuring specific items for inventory, as contrasted with management of inventory as whole: and so on.

Distinctions Between Task Control and Management Control

Task control is distinguished from management control in the following ways:

- The management control system is basically of similar throughout the organization. Each type task requires a different task control system.

- In management control, managers interact with other managers; in task control either humans are not involved at all, or the interaction is between a manager and a nonmanager.

- In management control the focus is on organizational units called responsibility centers; in task control the focus is on specific tasks.

- Management control relates to activities that are not specified; task control relates to specified tasks.

- In management control the focus is equally on planning and on execution; in task control it is primarily on execution.

An essential characteristic of the process is that the "*standard*" against which actual performance is measured is consistent with the organization's strategies. *Exhibit 6-3* outlines differences among the three types of processes with respect to the nature of the problems that typically are addressed in each process and the types of decisions that are relevant for these problems.

As another way of explaining the differences among the three processes, *Exhibit 6-4* gives some examples of activities associated with each.

Strategic Control: A New Perspective

Most commentators would agree with the definition of strategic control offered by *Schendel and Hofer:*

"Strategic control focuses on the dual questions of whether:

(1) the strategy is being implemented as planned; and

(2) the results produced by the strategy are those intended."

This definition refers to the traditional review and feedback stages which constitutes the last step in the strategic management process. Normative models of the strategic management process have depicted it as including there primary stages: strategy formulation, strategy implementation, and strategy evaluation (control).

Strategy evaluations concerned primarily with traditional controls processes which involves the review and feedback of performance to determine if plans, strategies, and objectives are being achieved, with the resulting information being used to solve problems or take corrective actions.

Recent conceptual contributors to the strategic control literature have argued for anticipatory feedforward controls, that recognize a rapidly changing and uncertain external environment.

Schreyogg and Steinmann (1987) have made a preliminary effort, in developing new system to operate on a continuous basis, checking and critically evaluating assumptions, strategies and results. They refer to strategic control as "**the critical evaluation of plans, activities, and results, thereby providing information for the future action**".

Schreyogg and Steinmann based on the shortcomings of feedback-control. Two central characteristics if this feedback control is highly questionable for control purposes in strategic management:

(a) *feedback control is post-action control and*

(b) *standards are taken for granted.*

Schreyogg and Steinmann proposed an alternative to the classical feedback model of control: a 3-step model of strategic control which includes **premise control**, **implementation control**, and **strategic surveillance**. *Pearce and Robinson* extended this model and added a component "**special alert control**" to deal specifically with low probability, high impact threatening events.

The nature of these four strategic controls is summarized in *Figures*. Time (t) marks the point where strategy formulation starts. Premise control is established at the point in time of initial premising (t). From here on promise control accompanies all further selective steps of premising in planning and implementing the strategy. The strategic surveillance of emerging events parallels the strategic management process and runs continuously from time (t) through (t). When strategy implementation begins (t), the third control device, *implementation control* is put into action and run through the end of the planning cycle (t). Special alert controls are conducted over the entire planning cycle.

Promise Control

Planning premises/assumptions are established early on in the strategic planning process and act as a basis for formulating strategies.

"Premise control has been designed to check systematically and continuously whether or not the premises set during the planning and implementation process are still valid.

It involves the checking of environmental conditions. Premises are primarily concerned with two types of factors:

- **Environmental factors** (for example, inflation, technology, interest rates, regulation, and demographic/social changes).

- **Industry factors** (for example, competitors, suppliers, substitutes, and barriers to entry).

All premises may not require the same amount of control. Therefore, managers must select those premises and variables that:

(a) are likely to change and

(b) would a major impact on the company and its strategy if the did.

Implementation Control

Strategic implantation control provides an additional source of feedforward information.

"Implementation control is designed to assess whether the overall strategy should be changed in light of unfolding events and results associated with incremental steps and actions that implement the overall strategy."

Strategic implementation control does not replace operational control. Unlike operations control, strategic implementation control continuously questions the basic direction of the strategy. The two basis types of implementation control are:

1. **Monitoring strategic thrusts (new or key strategic programs).** Two approaches are useful in enacting implementation controls focused on monitoring strategic thrusts: (a) one way is to agree early in the planning process on which thrusts are critical factors in the success of the strategy or of that thrust; (b) the second approach is to use stop/go assessments linked to a series of meaningful thresholds (time, costs, research and development, success, *etc.*) associated with particular thrusts.

2. **Milestone Reviews.** Milestones are significant points in the development of a programme, such as points where large commitments of resources must be made. A milestone review usually involves a full-scale reassessment of the strategy and the advisability of continuing or refocusing the direction of the company. In order to control the current strategy, must be provided in strategic plans.

Strategic Surveillance

Compared to premise control and implementation control, strategic surveillance is designed to be a relatively unfocused, open, and broad search activity.

Strategic surveillance is designed to monitor a broad range of events inside and outside the company that are likely to threaten the course of the firm's strategy.

The basic idea behind strategic surveillance is that some form of general monitoring of multiple information sources should be encouraged, with the specific intent being the opportunity to uncover important yet unanticipated information.

Strategic surveillance appears to be similar in some way to *"environmental scanning."* The rationale, however, is different. Environmental, scanning usually is seen as *part* of the chronological *planning cycle* devoted to generating information for the new plan.

By way of contrast, strategic surveillance is designed to safeguard the *established strategy*on a continuous basis.

Special Alert Control

Another type of strategic control is a special alert control.

"A special alert control is the need to thoroughly, and often rapidly, reconsider the firm's basis strategy based on a sudden, unexpected event."

The analysts of recent corporate history are full of such potentially high impact surprises (*i.e.*, natural disasters, chemical spills, plane crashes, product defects, hostile takeovers *etc.*).

While Pearce and Robinson suggest that special alert control be performed only during strategy implementation, Preble recommends that because special alert controls are really a subset of strategic surveillance that they be conducted throughout the entire strategic management process.

The characteristics of each control component are detailed in *Tables*, including the component's purpose, mechanism used to implement it, the procedure to be followed, degree of focusing, information sources, and organizational/personnel to be utilized.

Strategic Control Process

Although control systems must be tailored to specific situations, such systems generally follow the same basic process.

Regardless of the type or levels of control systems an organization needs, control may be depicted as a six-step feedback model):

1. **Determine what to control**. *What are the objectives the organization hopes to accomplish?*
2. **Set control standards**. *What are the targets and tolerances?*
3. **Measure performance**. *What are the actual standards?*
4. **Compare the performance the performance to the standards**. *How well does the actual match the plan?*

5. **Determine the reasons for the deviations**. *Are the deviations due to internal shortcomings or due to external changes beyond the control of the organization?*

6. **Take corrective action**. *Are corrections needed in internal activities to correct organizational shortcomings, or are changes needed in objectives due to external events?*

Feedback from evaluating the effectiveness of the strategy may influence many of other phases on the strategic management process.

A well-designed control system will usually include feedback of control information to the individual or group performing the controlled activity.

Simple feedback systems measure outputs of a process and feed into the system or the inputs of a system corrective actions to obtain desired outputs. The consequence of utilizing the feedback control systems is that the unsatisfactory performance continues until the malfunction is discovered. One technique for reducing the problems associated with feedback control systems is **feedforward control**. Feedforward systems monitor inputs into a process to ascertain whether the *inputs* are as planned; if they are not, the inputs, or perhaps the process, are changed in order to obtain desired results.

DETERMINE WHAT TO CONTROL

The first step in the control process is determining the major areas to control. Managers usually base their major controls on the organizational mission, goals and objectives developed during the planning process.

Managers must make choices because it is expensive and virtually impossible to control every aspect of the organization's activities. In deciding what to control, the organization must communicate through the actions of its executives that strategic control is a needed activity. Without top management's commitment to controlling activities, the control system could be useless.

Set Control Standards

The second step in the control process is establishing standards. A **control standards** is a target against which subsequent performance will be compared.

Standards are the criteria that enable managers to evaluate future, current, or past actions. They are measured in a variety of ways, including physical, quantitative, and qualitative terms. Five aspects of the performance can be managed and controlled: **quantity, quality, time cost, and behavior**. Each aspect of control may need additional categorizing.

An organization must identify the targets, determine the tolerances those targets, and specify the timing of consistent with the organization's goals defined in the first step of determining what to control. For example, standards might indicate how well a product is made or how effectively a service is to be delivered.

Standards may also reflect specific activities or behaviors that are necessary to achieve organizational goals. Goals are translated into performance standards by making them measurable. An organizational goal to increase market share, for example, may be translated into a top-management performance standard to increase market share by 10 percent within a twelve-month period. Helpful measures of strategic performance include: sales (total, and by division, product category, and region), sales growth, net profits, return on sales, assets, equity, and investment cost of sales, cash flow, market share, product quality, valued added, and employees productivity.

Quantification of the objective standard is sometimes difficult. For example, consider the goal of product leadership. An organization compares its product with those of competitors and determines the extent to which it pioneers in the introduction of basis product and product improvements. Such standards may exist even though they are not formally and explicitly stated.

Setting the timing associated with the standards is also a problem for many organizations. It is not unusual for short-term objectives to be met at the expense of long-term objectives.

Management must develop standards in all performance areas touched on by established organizational goals. The various forms standards are depend on what is being measured and on the managerial level responsible for taking corrective action.

Commonly uses as an example, the following eight types of standards have been set by **General Electric** :

- **Profitability standards** : These standards indicate how much profit General Electric would like to make in a given time period.
- **Market position standards** : These standards indicate the percentage of total product market that company would like to win from competitors.
- **Productivity standards** : These production-oriented standards indicate various acceptable rates which final products should be generated within the organization.
- **Product leadership standards** : Product leadership standards indicate what levels of product innovation would make people view General Electric products as leaders in the market.
- **Personnel development standards** : Personnel development standards list acceptable of progress in this area.
- **Employee attitude standards** : These standards indicate attitudes that General Electric employees should adopt.
- **Public responsibility standards** : All organizations have certain obligations to society. General Electric's standards in this area indicate acceptable levels of activity within the organization directed toward living up to social responsibilities.

- Standards reflecting balance between short-range and long-range goals . Standards in this area indicate what the acceptable long- and short - range goals are and the relationship among them.

Critical Control Points and Standards. The principle of critical point control, one of the more important control principles, states: **"Effective control requires attention against plans"**. There are, however, no specific catalog of controls available to all managers because of the peculiarities of various enterprises and departments, the variety of products and services to be measured, and the innumerable planning programs to be followed.

MEASURE PERFORMANCE

Once standards are determined, the next step is measuring performance. The actual performance must be compared to the standards. In some work places, this phase may require only visual observation. In other situations, more precise determinations are needed. Many types of measurements taken for control purposes are based on some form of historical standard. These standards can be based on data derived from the **PIMS (profit impact of market strategy)** program, published information that is publicly available, ratings of product / service quality, innovation rates, and relative market shares standings. PIMS was developed by Professor Sidney Shoeffler of Harvard University in the 1960s.

Strategic control standards are based on the practice of **competitive benchmarking** - the process of measuring a firm's performance against that of the top performance in its industry.

The proliferation of computers tied into networks has made it possible for managers to obtain up-to-minute status reports on a variety of quantitative performance measures. Managers should be careful to observe and measure in accurately before taking corrective action.

Compare Performance to Standards

The comparing step determines the degree of variation between actual performance and standard. If the first two phases have been done well, the third phase of the controlling process - comparing performance with standards - should be straightforward. However, sometimes it is difficult to make the required comparisons (*e.g.*, behavioral standards).

Some deviations from the standard may be justified because of changes in environmental conditions, or other reasons.

Determine the Reasons for the Deviations

The fight step of the control process involves finding out: "why performance has deviated from the standards?" Causes of deviation can range from selected achieve organizational objectives. Particularly, the organization needs to ask if the deviations are due to internal shortcomings or external changes beyond the control of the organization.

A general checklist such as following can be helpful:

- Are the standards appropriate for the stated objective and strategies?
- Are the objectives and corresponding still appropriate in light of the current environmental situation?
- Are the strategies for achieving the objectives still appropriate in light of the current environmental situation?
- Are the firm's organizational structure, systems (*e.g.*, information), and resource support adequate for successfully implementing the strategies and therefore achieving the objectives?
- Are the activities being executed appropriate for achieving standard?

The locus of the cause, either internal or external, has different implications for the kinds of corrective action.

Take Corrective Action

The final step in the control process is determining the need for corrective action. Managers can choose among three courses of action:

1. they can do nothing
2. they can correct the actual performance; or
3. they can revise the standard.

Maintaining the status quo if preferable when performance essentially matches the standards. When standards are not met, managers must carefully assess the reasons why and take corrective action. Moreover, the need to check standards periodically to ensure that the standards and the associated performance measures are still relevant for the future.

The final phase of controlling process occurs when managers must decide action to take to correct performance when deviations occur. Corrective action depends on the discovery of deviations and the ability to take necessary action. Often the real cause of deviation must be found before corrective action can be taken. Causes of deviations can range from unrealistic objectives to the wrong strategy being selected achieve organizational objectives. Each cause requires a different corrective action. Not all deviations from external environmental threats or opportunities have progressed to the point a particular outcome is likely, corrective action may be necessary.

There are three choices of corrective action:

1. **Normal mode** - follow a routine, no crisis approach; this take more time
2. **As hoc crash mode** - saves time by speeding up the response process, geared to the problem ad hand.

3. **Preplanned crisis mode** - specifies a planned response in advance; this approach lowers the response time and increases the capacity for handling strategic surprises.

The below checklist suggest the following five general areas for corrective actions:

- **Revise the Standards.** It is entirely possible that the standards are not in line with objectives and strategies selected. Changing an established standard usually is necessary if the standards were set too high or to low are the outset. In such cases it's the standard that needs corrective attention not the performance.

- **Revise the Objective.** Some deviations from the standard may by justified because of changes in environmental conditions, or other reasons. In these circumstances, adjusting the objectives can y much more logical and sensible then adjusting performance.

- **Revise the Strategies.** Deciding on internal changes and taking corrective action may involve changes in strategy. A strategy that was originally appropriate can become inappropriate during a period because of environmental shifts.

- **Revise the Structure, System or Support.** The performance deviation may by caused by an inadequate organizational structure, systems, or resource support. Each of these factors is discussed elsewhere in this chapter, or other part of this thesis.

- **Revise Activity.** The most common adjustment involves additional coaching by management, additional training, more positive incentives, more negative incentives, improved scheduling, compensation practices, training programs, the redesign of jobs or the replacement of personnel.

Managers can also attempt to influence events or trends external to itself through advertising or other public awareness programs. In such case, the changes should be made only after the most intense scrutiny.

Management must remember that adjustments in any of the above areas may require adjustments in one or more of the other factors. For example, adjusting the objectives is likely to require different strategies, standards, resources, activities, and perhaps organizational structure and systems.

Strategy Audits

In order to better understand what strategic control performance measures are and how a manager can take such measurements, we need to introduce two important topics:

(1) strategic audits and

(2) strategic audit measurement methods.

Strategic Audits

A strategic audit is an examination and evaluation of areas affected by the operation of a strategic management process within an organization. A strategy audit may be needed under the following conditions:

- Performance indicators show that a strategy is not working or is producing negative side effects.
- High-priority items in the strategic plan are not being accomplished.
- A shift or change occurs in the external environment.
- Management wishes: (1) to fine-tune a successful strategy and (2) to ensure that a strategy that has worked in the past continues to be in tune with subtle internal or external changes that may have occurred.

Assessing the Firm's Operational and Strategic Health

To aid in control, firms will occasionally perform audits to ensure that certain aspects of their operations are in order. Such audit may include **operational audits** (assessing the firm's operating health) and **strategic audits**(assessing the firm's strategic health).

MEASURES OF ORGANIZATIONAL HEALTH

Measures or indicators of a firm's current operating and strategic health are shown in Tables. As the tables show, to assess a firm's current operating health, short-term financial, market, technological, and production position are used, while current strategic health is based on strategic market position, technological position, production capabilities, and financial health.

Strategic Audit Measurement Methods

There are several generally accepted methods for measuring organizational performance. One way for categorizing these methods divides into the distinct types: **qualitative and quantitative**. However, a few methods do not fall neatly into one or other of these categories but rather are a combination of both types.

Qualitative Organizational Measurements

There is no universally endorsed list of critical questions designed to reflect important facets of organizational operations. However, several that might be useful to the practicing managers are presented below.

Sample Questions to be asked for Qualitative Organizational Measurement:

- Are the financial policies with respect to investment… dividends and financing consistent with opportunities likely to be available?
- Has the company defined the market segments in which it intends to operate sufficiently specifically with respect to both product lines and

market segments? Has it clearly defined the key capabilities needed for success?

- Does the company have a viable plan for developing a significant and defensible superiority over competition with respect to these capabilities?
- Will the business segments in which the company operates provide adequate opportunities for achieving corporate objectives? Do they appear as attractive as to make it likely that an excessive amount of investment will be drawn to the market from other companies? Is adequate provision being made to develop attractive new investment opportunities?
- Are the management, financial, technical and other resources of the company really adequate to justify an expectation of maintaining superiority over competition in the key areas of capability?
- Does the company have operations in which it is not reasonable to expect to be more capable than competition? If so, can the board expect them to generate adequate returns on invested capital? Is there any justification for investing further in such operations, even just to maintain them?
- Has the company selected business that can reinforce each other by contributing jointly to the development of key capabilities? Or are there competitors that have combinations of operations which provide them with an opportunity to gain superiority in the key resource areas? Can the company's scope of operations be revised so as to improve its position vis-à-vis competition?
- To the extent that operations are diversified, has the company recognized and provided for the special management and control systems required?

The Evaluation of Corporate Strategy

Each organization has its own approach to evaluation. There are not absolute answers as to the proper evaluation standards. However, there are three basic questions to ask in strategy evaluation:

1. Is the existing strategy any good?
2. Will the existing strategy be good in the future?
3. Is there a need to change a strategy?

The first question may need additional detailing to indicate whether the current strategy is useful and beneficial to the organization.

Seymour Tilles has written a classic article on the qualitative assessment of organizational performance. This article serves several particular questions to be asked for evaluation. These questions are:

1. Is the strategy internally consistent? Internal consistency refers to the cumulative impact of various strategies on the organizations. According to Tilles, a strategy must be judged not only in relationships to other strategies.

2. Is organizations strategy consistent with its environment? An important test of strategy is whether the chosen strategy in consistent with environment (constituent demands, competition, economy, product / industry life cycle, suppliers, customers) - whether the really make sense with respect to what is going on outside.

3. Is the strategy appropriate in view of available resources? Resources are those things that company is or has and that help it to achieve its corporate objectives. Included are money, competence, facilities and other. Without appropriate resources, organization simply cannot make strategic work.

4. Does the strategy involve an acceptable degree of risk? Strategy and resources, taken together, determine the degree of risk which the company is undertaken. Each company must determine the amount of risk it wishes to incur. This is a critical managerial choice. In attempting to assess the degree of risk associated with a particular strategy, management must assess such issues as the total amount of resources a strategy requires, the proportion of the organization's resources that a strategy will consume, and the amount of time that must be committed.

5. Does the strategy have an appropriate time horizon? A significant part of every strategy is the time horizon on which it is based. For example, a new product developed, a plant put on stream, a degree of market penetration, become significant strategic objectives only if accomplished by a certain time. Management must ensure that the time necessary to implement the strategy is consistent. Inconsistency between these two variables can make it impossible to reach goals in a satisfactory way.

6. Is the strategy workable?

E. P. Learned and others, building on the Tilles model, suggest that the following are also proper evaluative questions:

7. Is the strategy identifiable? Has it been clearly and consistently identified and are people aware of it?

8. Is the strategy appropriate to the personal values and aspirations of key managers?

9. Does strategy constitute a clear stimulus to organizational effort and commitment?

10. Is the strategy socially responsible?

11. Are there early indications of the responsiveness of markets and market segments to the strategy?

J. Argenti adds:

12. Does the strategy rely on weakness or do anything to reduce them?

13. Does the strategy exploit major opportunities?

14. Does it avoid, reduce, or mitigate the major threats? If not, are there adequate contingency plans?

All these questions can by applied as the strategy progresses through its various stages, including implementation. The answers can provide guidelines as to how the strategy should be altered or changed.

The second basic question "Will the existing strategy be good in the future?" seeks to ascertain if the strategy would continue to satisfy the firm's objective in the future. The answer to this is based upon unforeseeable changes in the organization's environment or resources, or changes in its mission, goals, or objectives.

The answer to the third question "Is there a need to change the strategy?" will provide direction toward a strategy formation task.

Qualitative measurements methods can be very useful, but their application involves significant amounts of human judgment. Thus, conclusions based on such methods must be drawn carefully.

Quantitative Organizational Measurements

Quantitative measurements provide information and insight as to how well an organization is accomplishing its goods and objectives. In attempting to evaluate the effectiveness of corporate strategy quantitatively, we can see how the firm has done compared wit its own history, or compared with its competitors.

Many quantitative measures may be developed to determine performance results. These standards expressed in quantitative terms include:

1. Sales (growth of sales)
2. Net profit
3. Dividend returns
4. Return on equity
5. Return on investment
6. Return on capital
7. Marker share
8. Earnings per share

The list is long and many other factors could be included. The objective of all of these endeavors is financial control.

But financial control is only part of the total strategic management control process. Much of the activity affects financial performance in non financial nature. This include consideration of labor efficiency and productivity; production quantity turnover, and tardiness; on a very limited basis, human resources accounting and personnel satisfaction measures; more commonly, management by objectives systems; social analysis; operational audits of any functional, divisional, or staff component, distribution cost and efficiency; management audits modeling; and so forth.

The list is almost endless and there is no time to discuss each item here.

Which factors should be used? Establishing the standards and tolerance limit is not as easy as we might expect. Managers need to first define the critical success factors - the factors which are most important to the strategy and being successful in the business. Most of these measures are internal. But objective assessments can also be made by comparing the firm's results of similar firms.

Below we present a set of worthwhile guidelines that managers might follow in designing and implementing more comprehensive strategic audits.

A strategic audit is conducted in three phases: diagnosis to identify how, where, and in what priority in-depth analyses need to be made; focused analysis; and generation and testing recommendation. Objectivity and the ability to ask critical, probing questions are key requirements for conducting a strategic audit.

Phase One: Diagnosis

The diagnostic phase includes the flowing tasks:

1. Review key document such as:

    o Strategic plan

    o Business or operational plans

    o Organizational arrangements

    o Major policies governing matters such as resource allocation and performance measurement.

2. Review financial, market, and operational performance against benchmarks and industry norms to identify jet variances and emerging trends.

3. Gain an understanding of:

    o Principal roles, responsibilities, and reporting relationships.

    o Decision - making processes and major decisions made.

    o Resources, including physical facilities, capital, management, technology.

    o Interrelationships between functional staffs and business or operating units.

4. Identify strategic implications of strategy for organization structure, behavior patterns, systems, and processes.

    o Define interrelationships and linkages to strategy.

5. Determine internal and external perspectives.

    o Survey the attitudes and perceptions of senior and middle managers and other key employees to assess the extent to which these are consistent with the strategic direction of the firm. One way to accomplish this task is through carefully focused interviews and / or questionnaires, wherein employees are asked to identify and make trade-offs among the objectives and variables they consider most important.

    o Interview a carefully selected sample of customers and prospective customers and other key external sources to gain understanding of how the company is viewed.

6. Identify aspects of the strategy that are working well. Formulate hypotheses regarding problems and opportunities for improvement based on the findings above. Define how and in what order each should be pursued.

Phase Two: Focused analysis

1. Test the hypotheses concerning problems and opportunities for improvement through analysis of specific issues.

    o Identify interrelationships and dependencies among components of the strategic system.

2. Formulate conclusions as to weaknesses in strategy formulation, implementation deficiencies, or interactions between the two.

Phase Three: Recommendations

1. Develop alternative solutions to problems and ways of capitalizing on opportunities.

    o Test [these alternatives] in light of their resource requirements, risk, rewards, priorities, and other applicable measures.

2. Develop specific recommendations.

    o Develop an integrated, measurable, and time - phased action plan to improve strategic results.

Chapter 8

MONITORING AND CONTROL INTERFACE BASED ON VIRTUAL SENSORS

Ricardo F. Escobar[1], Manuel Adam-Medina[1], Carlos D. García-Beltrán[1],
Víctor H. Olivares-Peregrino[1,*], David Juárez-Romero[2] and Gerardo V. Guerrero-Ramírez[1]

[1] Centro Nacional de Investigación y Desarrollo Tecnológico, Tecnológico Nacional de México, Int. Internado Palmira S/N, Palmira, C.P. 62490 Cuernavaca, Morelos, Mexico; E-Mails: esjiri@cenidet.edu.mx (R.F.E.); adam@cenidet.edu.mx (M.A.-M.); cgarcia@cenidet.edu.mx (C.D.G.-B.); gerardog@cenidet.edu.mx (G.V.G.-R.)
[2] Centro de Investigación en Ingeniería y Ciencias Aplicadas, Universidad Autónoma del Estado de Morelos, Av. Universidad 1001, Col. Chamilpa, C.P. 62209 Cuernavaca, Morelos, Mexico; E-Mail: djuarezr7@gmail.com

* Author to whom correspondence should be addressed; E-Mail: olivares@cenidet.edu.mx; Tel.: +52-777-362-7770; Fax: +52-777-362-7795.

External Editor: Vittorio M.N. Passaro

ABSTRACT

In this article, a toolbox based on a monitoring and control interface (MCI) is presented and applied in a heat exchanger. The MCI was programed in order to realize sensor fault detection and isolation and fault tolerance using virtual sensors. The virtual sensors were designed from model-based high-gain observers. To develop the control task, different kinds of control laws were included in the monitoring and control interface. These control laws are PID, MPC and a non-linear model-based control law. The MCI helps to maintain the heat exchanger under operation, even if a temperature outlet sensor fault occurs; in the case of outlet temperature sensor failure, the MCI will display an alarm. The monitoring and control interface is used as a practical tool to support electronic engineering students with heat transfer and control concepts to be applied in a double-pipe heat exchanger pilot plant. The method aims to teach the students through the observation and manipulation of the main variables of the process and by the interaction with the monitoring and control interface (MCI) developed in Lab-VIEW©. The MCI provides the electronic engineering students with the knowledge of heat exchanger behavior, since the interface is provided with a thermodynamic model that approximates the temperatures and the physical properties of the fluid (density and heat capacity). An advantage of the interface is the easy manipulation of the actuator for an automatic or manual operation. Another advantage of the monitoring and control interface is that all algorithms can be manipulated and modified by the users.

Keywords

Monitoring and control interface; virtual sensors; interactive educational workstation; heat exchanger

1. INTRODUCTION

It has been observed that due to the academic training received by electronic engineering students, it is difficult for them to understand heat transfer concepts and how to deal with cases in which a failure occurs. Therefore, at the National Center for Research and Technological Development (CENIDET (Centro Nacional de Investigación y Desarrollo Tecnológico)), a monitoring and control interface (MCI) for a double-pipe heat exchanger was developed. The MCI is orientated to provide support for the electronic engineering students, and so, the MCI has two main purposes. On the one hand, this system provides the students with the knowledge of heat exchanger behavior. The MCI shows on-line the physical property behavior of the fluid in the double pipe heat exchanger ρ (density) and C_p (heat capacity). Additionally, the MCI shows on-line the value of the convective heat transfer coefficients (h_o, hi) and the global heat transfer coefficient (U), which are calculated by a thermodynamic model. On the other hand, the information obtained through the thermodynamic model is used by a nonlinear model that describes the temperature dynamics in the double pipe heat exchanger. A fault-tolerant system-based model was implemented as a practical example. The MCI allows manipulating the system actuators in order to perform control tasks. The interface can be operated in open loop or in closed loop. The main objective of the monitoring and control interface is to provide the students with educational workstation with practice in the control area. This work represents the effort of the lecturers who developed the educational workstation based on their own research, which is implemented in the heat exchanger pilot plant.

The main objective of the industry is to keep processes operating adequately; nevertheless, nowadays, the industry demands constant monitoring of the systems in order to improve the reliability and safety of the involved processes. Thus, monitoring systems must be efficient and capable of supervising the main variables of the process and to implement control tasks in order to enhance user capabilities. Industrial processes, such as mining, chemical, water treatment, among others, use complex systems that operate in different regimes, which make them difficult to control. Therefore, this work was developed in order to prepare electronic engineering students (EES) for these challenges and to give them multidisciplinary training.

Heat exchangers are widely used in power production, process, chemical and food industries, electronic, environmental engineering, waste heat recovery, manufacturing industries, space heating, refrigeration, air conditioning, chemical plants, petrochemical plants, petroleum refineries and natural gas processing. Therefore, it is extremely important for EES to learn the techniques of process control focused on heat exchangers.

Mathematical algorithms are becoming widely used to estimate unmeasurable variables. The available mathematical algorithms measure data to reconstruct unmeasurable variables or parameters of a system or a process. These kinds of systems are usually known as virtual sensors or soft sensors. Some applications of virtual sensors are: the monitoring of unknown variables or parameters of a mechanical system or a process [1–4]; the design of fault detection systems [5,6]; for fault detection isolation systems [7,8]; or for fault-tolerant control systems based on analytical redundancy [9]. In [10], a soft sensor is proposed using neural models in order to improve the product quality monitoring and control in a refinery. The advantages of the virtual sensors lie in that their implementation is easy, and in comparison with a hardware sensor, the cost is lower. In [11], the design and implementation of an observer-based soft sensor (virtual sensor) for a heat exchanger is presented.

The authors in [12] consider that the user interface is particularly important for educational software. There are some educational works concerning the heat exchanger [13,14] and thermodynamic systems [15,16], where the software is a teaching tool. The authors in [17] believe that learning can be enhanced by integrating the theoretical abstraction of textbooks with the tactile nature of the lab and plant.

Nowadays, the technological resources at universities and research departments allow one to develop sophisticated software with the aim of simulating the real conditions of the process. Most of the works in educational process are geared towards simulation [17–19]. However, developing practice or research in a pilot plant is more significant for the students, since they have the responsibility for the equipment and the process. Therefore, in this work, a monitoring and control interface for a heat exchanger pilot plant is presented.

2. THE HEAT EXCHANGER AND ITS INSTRUMENTATION

A double pipe heat exchanger is a device where two fluids exchange heat. The double pipe heat exchanger is formed by two circular concentric pipes. The heat exchanger can be operated in parallel (the fluids flow in the same direction through the tube side and the shell side) or in counter-current flow (the fluids flow in opposite directions). In this work, the counter-current flow was used, as is shown in Figure 1.

The main variables involved in this process are:

- The inlet temperature of the fluid in the hot stream, T_{hi}.
- The inlet temperature of the fluid in the cold stream, T_{ci}.
- The outlet temperature of the fluid in the hot stream, T_{ho}.
- The outlet temperature of the fluid in the cold stream, T_{co}.
- The hot water flow, W_{vh}.
- The cold water flow, W_{vc}.

The double-pipe heat exchanger pilot plant produced by Didatec Technologies is located at the Process Control Laboratory of CENIDET (see Figure 2). The plant operates as a water-cooling process. In this heat exchanger, there is no change in the phase of any interacting fluid.

The pilot plant is provided with the following instrumentation:

- T_{ci} and T_{ho} are measured by two RTDPt-100 with four wires.
- T_{co} and T_{hi} are measured by two RTD Pt-100 with three wires.
- W_{vc} and W_{vh} are measured in two Platon variable section flow meters, and there is no digital signal from the flow meters.

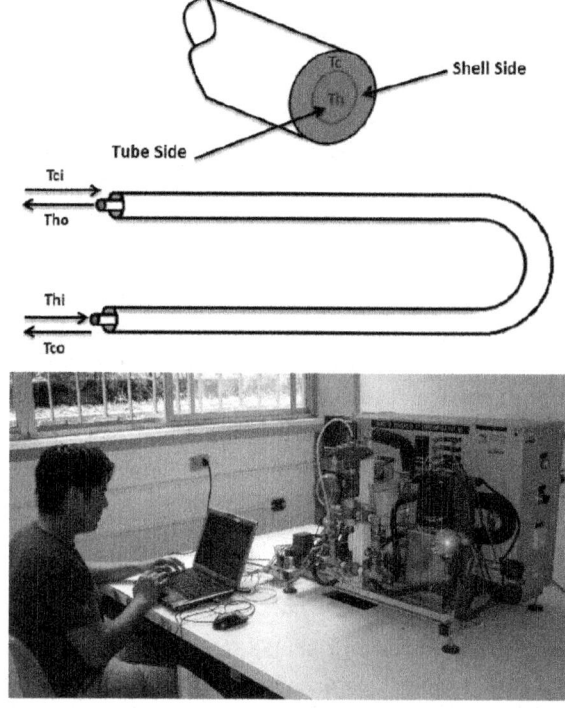

Figure 1. Countercurrent double-pipe heat exchanger.

Table 1. RTDPt-100 values.

RTD	≤±0.2 °C	≤±0.01 °C/°C
Lin.R	≤±0.1 Ω	≤±0.01 mΩ/°C
Volt	≤±10 µV	≤±1 mµ/°C
Temperature range RTD 3W		−40 °C to 85 °C
Temperature range RTD 4W		0 °C to 100 °C
Effect of supply voltage variation		<0.005% of span/VDC
Vibration		IEC 60068-2-6 Test FC
Max. wire size		1×1.5 mm^2 stranded wire
Dimensions		44×20.2 mm
Weight		50 g
Max offset		50% of selec.max value
Cable resistance per wire (max)		5 Ω
Output sensor current		0.2...0.4 mA
Effect of sensor cable resistance (3-/4-wire)		<0.002 Ω/Ω

The basic values of the Engelhard Pyro-Control Pt-100 temperature transmitter are given in Table 1. The instrumentation of the plant is sufficient to perform full variable monitoring, but not enough for automatic control tasks, since the flow rate cannot be recorded by the acquisition system. In the Didatec Technologies pilot-plant, the cold and hot water streams W_{vc}, $W_v h$ are not measured by a digital transmitter (W_{vc} is the manipulated variable, and $W_v h$ is taken as a constant). Furthermore, the system is not equipped with a human interface to allow the students to visualize, adequately, the data provided by the mentioned instruments and/or to control the plant actuators. Therefore, the students must be near the plant to visualize the measurements of the desired variables, which limits the user functions.

The goal of this work is to design a monitoring and control interface for the pilot plant in order to teach automatic control and heat transfer concepts. To develop the monitoring and control interface, an adequate acquisition system is required to transfer the plant's data to a computer, so it is necessary to integrate mathematic algorithms to estimate variables, such as temperature and flow rate.

Heat Exchanger Modeling

In order to show the behavior of the main process variables in the heat transfer process and to develop a control system, a dynamic model was implemented. Furthermore, the algorithm model has implemented a thermodynamic model, which evaluates the density, heat capacity, convective heat transfer coefficients (hi, h_o) and the global heat transfer coefficient; these values are displayed in a separate graphic.

The model was developed considering the follows conditions:

- A1. The flow is constant in the shell side.
- A2. The water physical properties are evaluated as functions of the temperature by empiric correlations.

- A3. The convective coefficient is dependent on the flow and temperature, and it varies with it.
- A4. There is no heat transfer between the outer pipe and the environment.
- A5. There is no energy accumulation in the pipe walls.
- A6. The system inputs are measurable.

$$
\begin{aligned}
\frac{T_{co}}{dt} &= \frac{W_{vc}}{V_{lc}}(T_{ci}(t) - T_{co}(t)) + \left(\frac{UA_o}{C_{p_c}\rho_c V_{lc}}\right)(T_{ho}(t) - T_{co}(t)) \\
\frac{T_{ho}}{dt} &= \frac{W_{vh}}{V_{lh}}(T_{hi}(t) - T_{ho}(t)) + \left(\frac{UA_i}{C_{p_h}\rho_h V_{lh}}\right)(T_{co}(t) - T_{ho}(t))
\end{aligned}
\tag{1}
$$

where A_o and Ai are the shell area and the tube side area, respectively. The thermodynamic model of the double-pipe heat exchanger was presented in [20].

3. THE DATA ACQUISITION SYSTEM

The programming software selected to develop the monitoring interface is LabVIEW©, due to it being provided with algorithms to perform easily the remote monitoring tasks and to monitor the control systems, as well as having a user-friendly graphical interface, which is why a National Instruments (NI©) acquisition card was selected to communicate the heat-exchanger pilot plant information to the computer.

The acquisition card used in this work is the NI USB-6008. It was selected because it is USB-compatible and has eight (12 bits) analog inputs, sufficient for acquiring all of the variables involved in the heat exchanger process; and it has two (12 bits) analog outputs used to send the data required to control the system actuators, as well as 12 digital I/O lines and a counter. The acquisition card is shown in Figure 3.

An advantages of this card is that it can be easily connected to the pilot plant, using a simple signal converter made by the students. The signal converter is used to convert the current signal (sensors signal) to the voltage signal (received by the DAQand then sending it to the computer), which is the input signal of the data acquisition. The connections of the signal converter are shown in Figure 4.

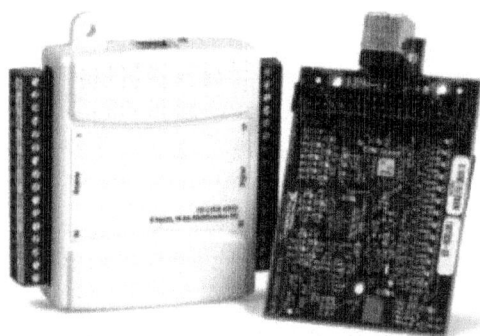

Figure 3. The selected National Instruments (NI) acquisition card.

Figure 4. Connections and conditioning stage.

4. DESIGN OF THE MONITORING AND INTERFACE

The designed interface is capable of performing three different tasks:

- Monitoring the inlet and outlet temperatures.
- Estimating the outlet temperatures, by using a heat exchanger model or by state observers to reconstruct a state variable from only one measure (T_{co} or T_{ho}) in order to implement a fault diagnosis scheme.
- Detecting and isolating sensor failures in the process with the aim of controlling the temperature, even if a failure occurs.

These tasks are executed simultaneously, as can be seen in Figure 5.

The main program of the MCI system has a hierarchic design, which is shown in Figure 5. It provides features to select the subprogram to perform. Figure 5 shows the hierarchic design of the MCI.

Figure 5. Hierarchic design.

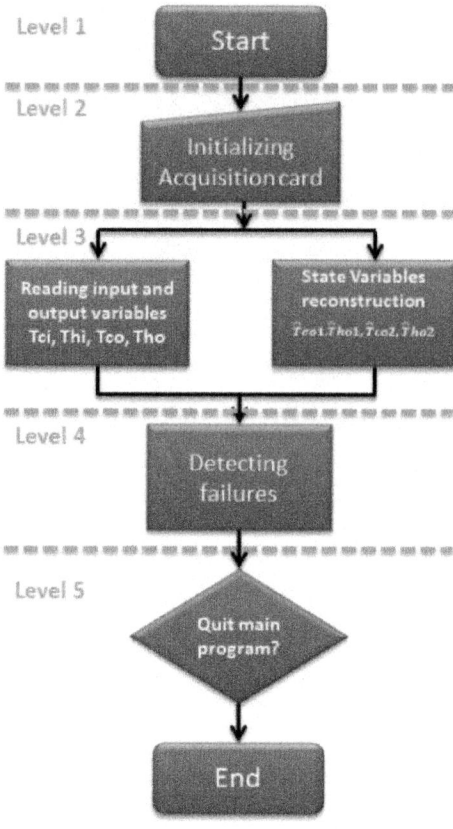

Level 1: Initialization of the main program. LabVIEW© has a running button to start the main program. Level 2: Initialization of the data acquisition. In the MCI, there is an acquisition button. Level 3: Signal reading and temperature estimation. The four temperature signals are read by the program, and these temperatures can be stored by a data store button, which is in the MCI. At the same level, there is the estimation of the temperatures, and at this level, a model and a bank of observers (virtual sensors) are programmed. The MCI has two buttons to request the model estimation and the virtual sensors estimation.

Level 4: The fault detection system, where the failure sensor could be replaced in case of failures. Level 5: A quit button, which can be pushed to end the MCI.

An intuitive interface is required in order to perform the monitoring and control tasks in the heat exchanger pilot plant. Due this, the interface requires an adequate communication with the pilot plant. An acquisition card developed by NI© is used to communicate data from/to the personal computer, which is why the LabVIEW© software was selected.

As can be seen in Figure 5, a routine is necessary in order to initialize the settings required to acquire and send data to/from the pilot plant and from/to the personal computer. This routine is configured by the NI© acquisition card, so it can be accessed by the computer using the indicated LabVIEW© program. The main flow diagrams for the listed tasks are explained below.

- Monitoring the temperatures plant: This algorithm reads a four-dimension data array, where the temperatures data are measured T_{ci}, T_{hi}, T_{co}, T_{ho} from the related sensors.

The monitoring program (represented by the flow diagram; Figure 6) developed in LabVIEW© allows the user to visualize the acquired data, through an on-line graphical interface. Different graphics are displayed for each temperature. The monitoring tasks are executed after the initializing routine automatically.

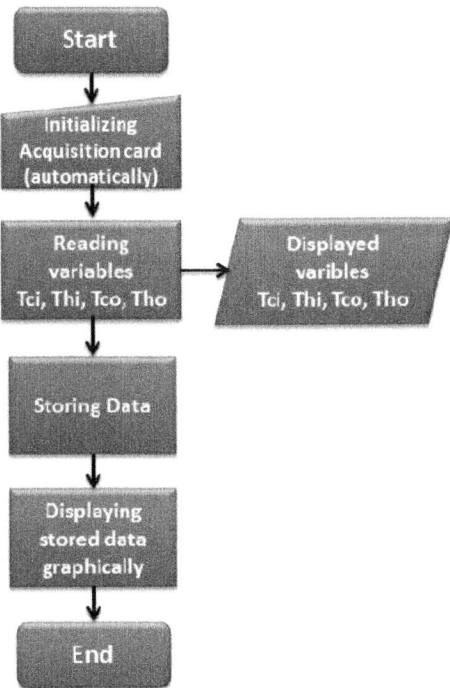

Figure 6. Variable monitoring.

- Estimating the outlet temperatures ($\hat{T}_{co}$ and $\hat{T}_{ho}$): This task executes an observer algorithm used to estimate both variables ($\hat{T}_{co}$ and $\hat{T}_{ho}$) by using state observers to reconstruct a state variable from only one measure (T_{co} or T_{ho}) in order to implement a fault diagnosis scheme. The observer algorithm uses the input variables and the available output measure. It performs the required control algorithm, which, based on a model of the pilot plant, estimates the value of the two output variables ($\hat{T}_{co}$ and $\hat{T}_{ho}$). The error is obtained by the comparison between the measured temperatures T_{co} and T_{ho} and the estimated temperatures $\hat{T}_{co}$ and $\hat{T}_{ho}$. The flow diagram is shown in Figure 7.

In the MCI, the temperature estimation (represented by the flow diagram shown in Figure 7) is executed by a button, which starts the estimation algorithms. The observer gain Equation (5) can be indicated in the interface or can be selected as a constant inside the program.

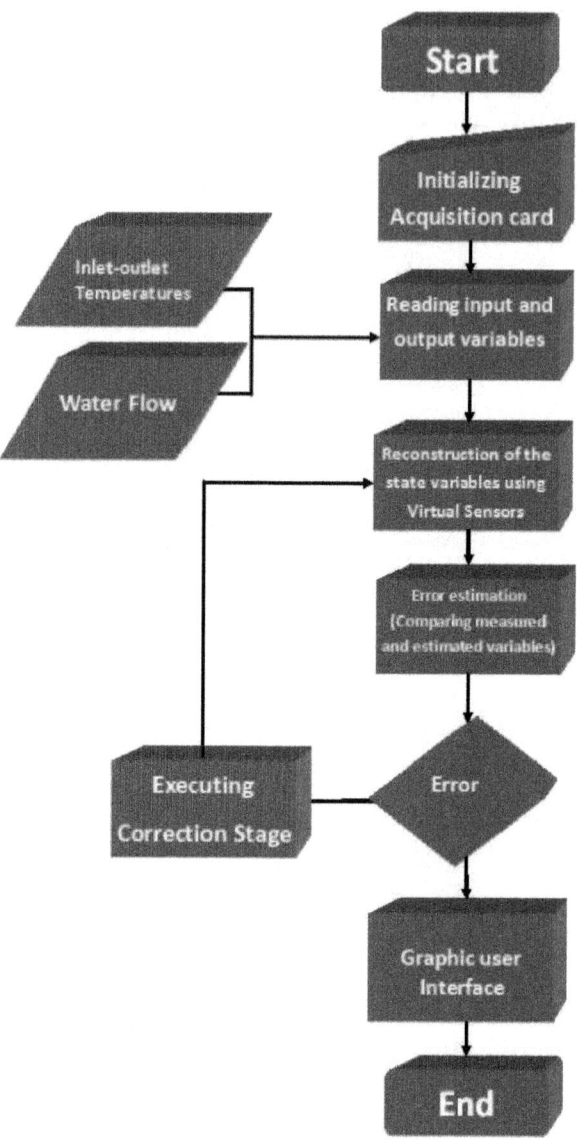

Figure 7. Estimating task.

- Detecting sensor failures: An observer-based algorithm is used to detect and isolate failures in the sensors. Two observers are required, one for each physical sensor. Observer 1 uses T_{ci}, T_{hi}, T_{co} as inputs, and Observer 2 uses T_{ci}, T_{hi}, T_{ho} as inputs. The observers estimate the temperatures in the cold and hot stream ($\hat{T}_{co}$, $\hat{T}_{ho}$). A comparison is done between each temperature estimation and its corresponding physical sensor measure in order to evaluate the error (so, there are four error evaluations Equations (6)–(9)). The error is compared with an error tolerance, which is set by the user. If the error is bigger than the error tolerance, then an alarm is displayed in the MCI, indicating that a failure exists.

Figure 8 shows the flow diagram that represents the fault detection realized by the virtual sensor or observer, and there is no interaction between the user and the interface when this task is running. In the case of a sensor failure, the MCI will display a green indicator.

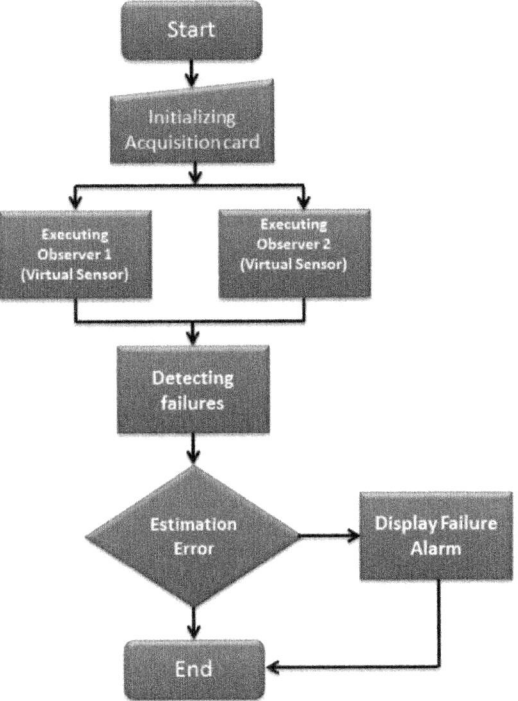

Figure 8. Failure detection function.

5. SUPERVISION AND CONTROL

5.1. Cold Water Flow Rate Solution

To monitor the heat exchanger control temperature using the cold water flow as the control input, it was necessary to characterize the control valve. In Figure 9, the valve characterization is shown, which presents hysteresis. Therefore, to send the control signal to the valve, it was convenient to use a Lagrange approximation (Equation (2)). The algorithm calculated the voltage necessary to control the valve according to the requested cold water flow (control input signal). It can be in open loop or in closed loop.

$$f(x_k) = P_n(x_k) \quad k = 0, 1, 2, \ldots, n$$

$$P_n(x) = f(x_0)L_{n,0}(x) + f(x_1)L_{n,1}(x) + \cdots + f(x_n)L_{n,n}(x) \tag{2}$$

$$L_{n,k}(x) = \frac{(X - X_0)(X - X_1)\cdots(X - X_{K+1})\cdots(X - X_n)}{(X_K - X_0)(X_K - X_1)\cdots(X_K - X_{K+1})\cdots(X - X_n)}$$

where x_k is the independent variable, $f(x_k)$ is the approximation of the independent variable, $L_{n,k}(x)$ is a polynomial base and $P_n(x)$ is the polynomial Lagrange interpolation.

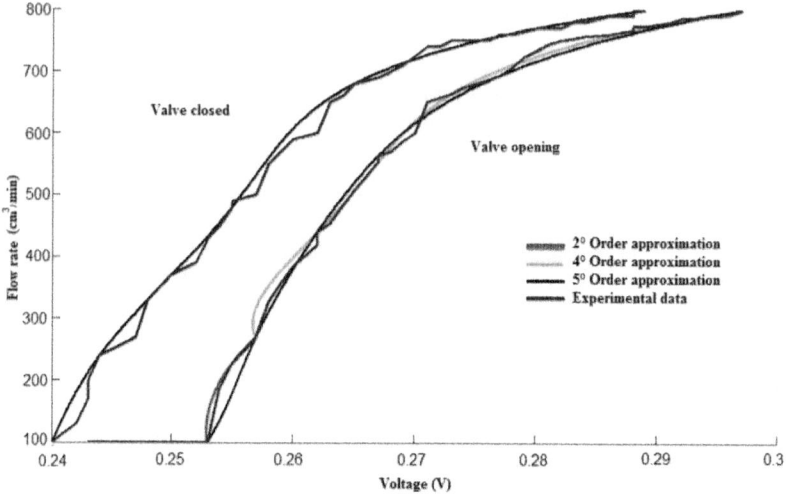

Figure 9. Valve characterization.

5.2. Fault Detection and Isolation-Based Model

To accomplish the sensor fault detection and isolation (FDI), analytical redundancy was employed. To develop this system, a bank of nonlinear high gain observers was implemented. The bank implementation was done, because the fault detection system was designed to detect two possible failures, in the cold temperature sensor (T_{co}) and in the hot temperature sensor (T_{ho}), so it is necessary to have two virtual sensors (or a bank of virtual sensors) in order to replace any of the two physical sensors. In the next section, a general explanation of the FDI system is given.

Fault Detection and Isolation Nonlinear Approach

To implement the FDI system, a bank of observers was implemented. The design of the FDI based on a bank of nonlinear observers was studied in [20]. Therefore, in the present work, the equations are presented in a general form.

5.3. Control-Affine Nonlinear System

A control-affine nonlinear system can be formulated as Equation (3).

$$\begin{cases} \dot{x}(t) = f(x(t)) + \sum_{i=1}^{m} g_i(x(t))u_i(t) \\ y(t) = h(x(t)) \end{cases} \tag{3}$$

where $x(t) \in \mathbb{R}^n$, $u_i(t) \in \mathbb{R}$, $i = 1, ..., m$, where m is the number of inputs, $y(t) \in \mathbb{R}^n$, $f(x(t)) \in \mathbb{R}^n$ and $g_i(x(t)) \in \mathbb{R}^n$, the last two are smooth vectors fields [20]. In these terms, if the model given in

Equation (1) can be formulated as Equation (3), then it could be possible to design a nonlinear observer Equation (4).

5.4. Nonlinear Observer

The high gain nonlinear observer is defined by Equation (4).

$$\begin{cases} \dot{\hat{x}}(t) = f(\hat{x}(t)) + \sum_{i=1}^{m} g_i(\hat{x}(t))\, u_i(t) - \left[\frac{\partial \Phi(\hat{x}(t))}{\partial \hat{x}} \right]^{-1} S_\theta^{-1} C^T [\hat{y}(t) - y(t)] \\ \hat{y}(t) = C\hat{x}(t) \end{cases} \tag{4}$$

where $\theta > 0$ is the tuning parameter of the observer $\dfrac{\partial \Phi(x(t))}{\partial x}$ and is the $n \times n$ Jacobian matrix of $\Phi(\hat{x}(t))$ and $\Phi(x(t)) = \Phi(x(t))\,|_{x(t)=\hat{x}(t)}$.

Considering a second order system, the matrix S_θ is:

$$S_\theta = \begin{bmatrix} \dfrac{1}{\theta} & -\dfrac{1}{\theta^2} \\ -\dfrac{1}{\theta^2} & \dfrac{2}{\theta^3} \end{bmatrix} \tag{5}$$

For the design of the FDI system, a bank of two observers is required.

5.5. Error Generation

The error evaluation between the physical sensors and the virtual sensors is obtained by Equations (6)–(9). An error evaluation is the difference between the measured temperature and the estimated temperature. Error evaluation from Observer 1:

$$r_{11} = \left| T_{co} - \hat{T}_{co_1} \right| \tag{6}$$

$$r_{12} = \left| T_{ho} - \hat{T}_{ho_1} \right| \tag{7}$$

Error evaluation from Observer 2:

$$r_{21} = |T_{co} - \hat{T}_{co2}| \tag{8}$$

$$r_{22} = |T_{ho} - \hat{T}_{ho2}| \tag{9}$$

where T_{ho} is the hot water outlet temperature in the pipe and T_{co} is the cooling water outlet temperature in the shell side. Therefore, the $\hat{T}_{co}$ and $\hat{T}_{ho}$ are estimated values, and the subscript number (1 or 2) refers to the observer number.

According to the authors in [20], the switching between the measure and by Observer 1 or 2 depends on the failure (output sensor of the hot water or output sensor of the cooling water, respectively). Thus, the following conditions are given in order to perform the switching.

$$T_{ho} = \begin{cases} T_{ho} & \text{if } r_{12} < \zeta \Rightarrow \text{Normal operation} \\ \hat{T}_{ho} & \text{if } r_{12} \geq \zeta \Rightarrow \text{A failure exists} \end{cases} \tag{10}$$

$$T_{co} = \begin{cases} T_{co} & \text{if } r_{21} < \zeta \Rightarrow \text{Normal operation} \\ \hat{T}_{co} & \text{if } r_{21} \geq \zeta \Rightarrow \text{A failure exists} \end{cases} \tag{11}$$

ζ is error tolerance, which is a constant value and represents the maximum value of the error between the estimated variable by the virtual sensor and the real sensor signal value. Once this value is reached by any residual (r_{12}, r_{21}), the real sensor signal is switched, and it is replaced by the virtual sensor. This error tolerance value is set by the user.

5.6. Selection of a Control Law

For academic purposes, the MCI has implemented different control laws, which are Proportional (P) (Equation (12)), Proportional-Integral (PI) (Equation (13)), Proportional-Integral-Derivative (PID) (Equation (14)), MPC (Equation (18)) and a nonlinear control law-based model (Equation (23)). With the implementation of these control laws, the students can practice different tuning methods. Furthermore, the MCI allows the interaction between the FDI system and the control law to ensure a fault-tolerant system.

$$P = K_p e(t) \tag{12}$$

$$PI = K_p + K_i \int_0^t e(t)dt \tag{13}$$

$$PID = K_p + K_i \int_0^t e(t)dt + K_d \frac{de}{dt} \tag{14}$$

To develop the prediction algorithm for the MPC, the state-space model given in Equation (15) was considered. The development of MPC is presented in detail in [21].

$$x_{k+1} = Ax_k + Bu_k \tag{15}$$
$$y_k + 1 = Cx_k + Du_k$$

where x is the state vector, y is the output vector and u is the input vector.

$$\begin{aligned} \frac{d\bar{x}}{dt} &= \mathbf{A}\bar{x} + \mathbf{B}\bar{u} \\ \bar{y} &= \mathbf{C}\bar{x} + \mathbf{D}\bar{u} \end{aligned} \tag{16}$$

In Equation (16), $\boldsymbol{x}$ denotes that the systems is linearized at one point. For linear systems of the form given in Equation (16), the development of the prediction model is simple:

$$\begin{aligned} \mathbf{x}_{k+n} &= \mathbf{A}^n\mathbf{x}_k + \mathbf{A}^{n-1}\mathbf{B}u_k + \mathbf{A}^{n-2}\mathbf{B}u_{k+1} + \cdots + \mathbf{B}u_{k+n-1} \\ y_{k+n} &= \mathbf{C}\left[\mathbf{A}^n\mathbf{x}_k + \mathbf{A}^{n-1}\mathbf{B}u_k + \mathbf{A}^{n-2}\mathbf{B}u_{k+1} + \cdots + \mathbf{B}u_{k+n-1}\right] \end{aligned} \tag{17}$$

To develop the control algorithm in the absence of input or output constraints, a generalized predictive control law is reduced to a linear feedback control strategy. State-space systems based on the predictive control law take the form:

$$\mathbf{u} - \mathbf{u}_{ss} = -k(\mathbf{x} - \mathbf{x}_{ss}) \tag{18}$$

A model-based nonlinear control law proposed in [22] was also implemented. The nonlinear control law Equation (23) is based on the nonlinear model of the heat exchanger Equation (1).

$$\ddot{e} = \ddot{y}_m - c\left(\frac{\partial f}{\partial x}\right)\dot{\mathbf{X}} \tag{19}$$

$$\frac{\partial f}{\partial x} = \left[\begin{array}{cc} \frac{1}{V_{lc}}\left(\dot{W}_{vc}(T_{ci} - T_{co}) - W_{vc}\right) + \frac{UA}{\rho_c Cp_c V_{lc}}\frac{\partial \Delta T ml}{\partial x_1} & \frac{UA}{\rho_c Cp_c V_{lc}}\frac{\partial \Delta T ml}{\partial x_2} \\ -\frac{UA}{\rho_h Cp_h V_{lh}}\frac{\partial \Delta T ml}{\partial x_1} & \frac{W_{vh}}{V_{lh}} - \frac{UA}{\rho_c Cp_c V_{lc}}\frac{\partial \Delta T ml}{\partial x_2} \end{array}\right] \tag{20}$$

$$\dot{e} = \lambda_m y_m - \lambda_m y_s \tag{21}$$

$$\ddot{e} = -k\left(e + \frac{1}{\tau_i}\int_0^t e\, dt + \tau_d\, \dot{e}\right) \tag{22}$$

$$W_{vh} = \frac{\frac{1}{V_{lc}}\left(\dot{W}_{vc}(T_{ci} - T_{co}) - W_{vc}\dot{T}_{co}\right) + \frac{UA}{\rho_c Cp_c V_{lc}}\left[\frac{\partial \Delta T ml}{\partial x_1}\frac{UA}{\rho_h Cp_h V_{lh}}\Delta T ml\right] + \ddot{e}}{\frac{UA}{\rho_c Cp_c V_{lc}}\frac{\partial \Delta T ml}{\partial x_2}(T_{hi} - T_{ho})} \tag{23}$$

The graphical user interface displays the parameters of the selected control law.

6. THE MONITORING AND CONTROL INTERFACE IMPLEMENTATION

The MCI was programed in order to realize sensors fault detection and isolation and fault tolerance using virtual sensors. The virtual sensors were designed from model-based high-gain observers. To develop the control task, different kinds of control laws were included in the monitoring and control interface. These control laws are PID, MPC and a nonlinear model-based control law. The MCI helps to maintain the heat exchanger under operation, even if a temperature outlet sensor fault occurs; in the case of outlet temperature sensor failure, the MCI displays an alarm. The monitoring and control interface is used as a practical tool to support electronic engineering students with heat transfer and control concepts to be applied in a double-pipe heat exchanger pilot plant. The method aims to teach the students through the observation and manipulation of the main variables of the process and by the interaction with the monitoring and control interface (MCI) developed in LabVIEW©. In this section, the monitoring and control interface is presented (Figure 10), which was developed in LabVIEW©, and some control algorithms of the MCI are shown.

The MCI allows the students to develop and implement approximation algorithms. The MCI is a tool that facilitates the interaction between the students and the process (heat exchanger). This tool allows students to understand the control algorithms that were programmed in the MCI. Furthermore, students can interact with the MCI in different ways, such as monitoring the states of the process, tuning the PID control law and developing control algorithms using the MATLAB programming language. At CENIDET, there are courses on MATLAB to train the students in programming skills. A common practice is to ask the students to develop a program for the heat exchanger's model using the MATLAB programming language, in order to realize a comparison with the real temperature signals acquired.

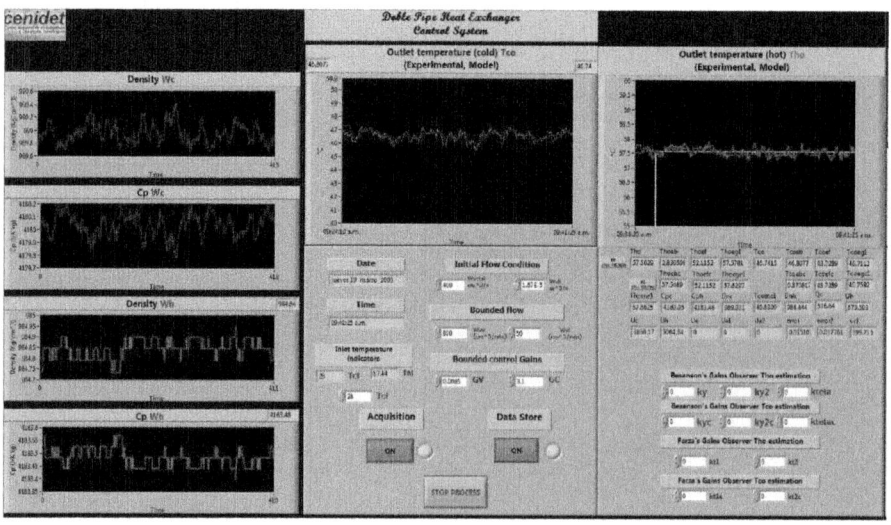

Figure 10. Main function graphical interface.

In Figure 10, the density and heat capacity of the fluid in both streams of the heat exchanger are shown (left-most section; tube side and shell side). In the upper center section of Figure 10, the cooling water temperature is shown (the red line shows the measured temperatures, and the blue line shows the estimated temperatures). In the right side of Figure 10, the hot water temperature is shown (the green line shows the measured temperatures, and the red line shows the estimated temperatures). The interface is intuitive for the users. Each section is adequately signposted, and it is continuously enhanced by the students according to their needs.

In Figure 11 (left side), the hot water temperature is shown (the green line shows measured temperature, and the red line shows the estimated temperatures). In the middle of the figure, the fault detection and control section are shown. As can be seen, the green light is lit, because of a fault in a hot water temperature sensor (total failure in the sensor). However, the system continues its operation, because the measured temperature is replaced by the estimated temperature (by *observer*$_1$), which does not depends on the T_{ho} measure for temperature. Other graphics are shown in this figure, such as the observer estimation (on the top), the control variable (cooling water flow) and the global heat transfer coefficient.

Figure 12 shows the program of the monitoring and control interface. In Figure 12, the data acquisition algorithm using LabVIEW© is presented. For this work, three temperatures were measured and acquired for the instrument $T_{hi'}$ T_{co} and T_{ho}.

At the top of the Figure 13, the error evaluation and the alarms section are shown. In the bottom of the figure (Figure 13), the nonlinear control law is shown, which was implemented for the closed loop; also, it has the Lagrange approximation to perform the closed loop control.

In Figure 14, the open loop control algorithm is shown for manual operation. The user can change the cold water flow manually through a slide manipulation, which is on the front panel of the MCI.

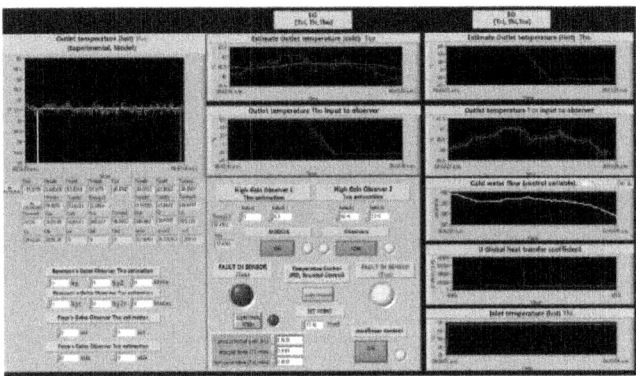

Figure 11. Fault-tolerant system.

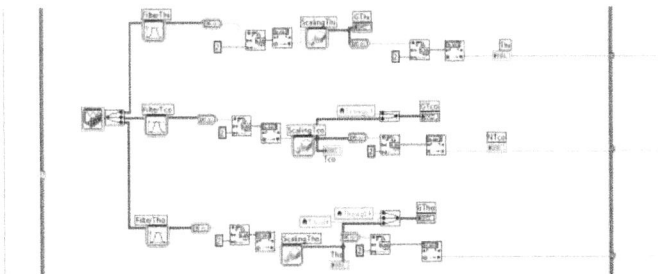

Figure 12. Acquisition algorithm using LabVIEW©.

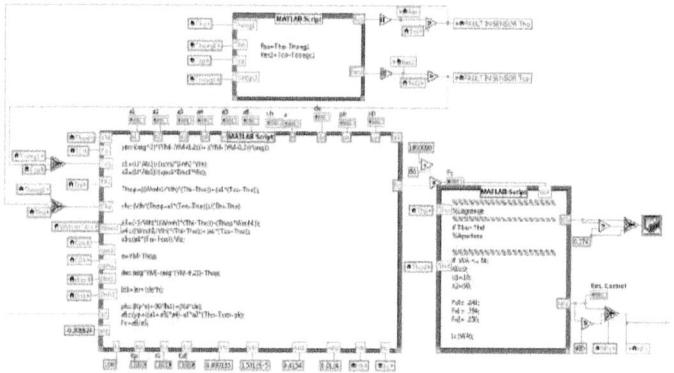

Figure 13. Fault detection algorithm.

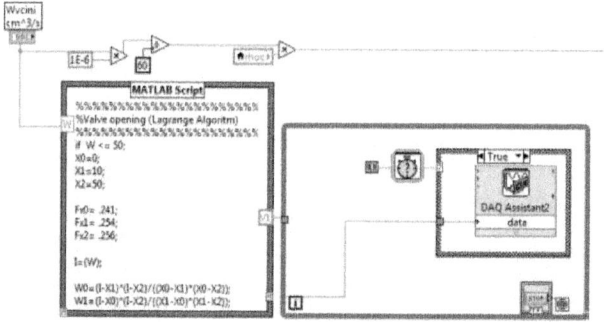

Figure 14. Open loop algorithm.

The figure shows the Lagrange approximation, which is used to calculate the required voltage for the control flow. Figure 15 shows the PID controller; also, it has the Lagrange approximation to perform the closed loop control.

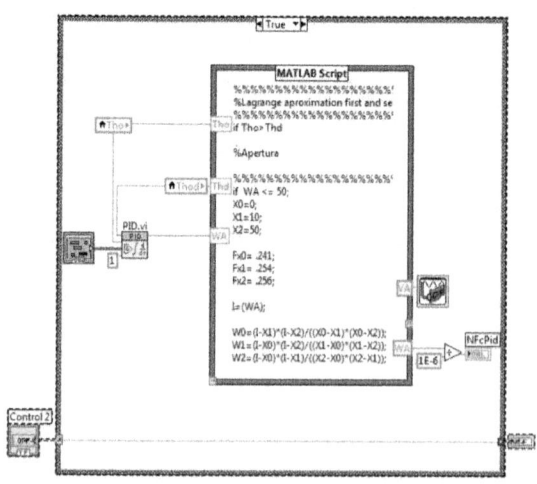

Figure 15. Closed loop algorithm.

7. RESULTS

With the monitoring control interface (MCI), the instructors for the control process, nonlinear control and classic control subjects have the support to develop their laboratory practices. Furthermore, most of the students can easily interact with the heat exchanger, even if they have little experience in heat exchange devices. The results obtained with the design of the MCI are: 100% of the EES that belong to the automatic control area have had the opportunity to practice at least once with the heat exchange devices during their stay at CENIDET. One student is developing his research on fault-tolerant control with an application in a heat exchanger. Another student is developing his research project on the design of state observers applied in a heat exchange process.

Figure 16 shows the opinion of the students about the MCI. In Figure 16, Sections A and B are the contents for the students that have made use of the MCI. In Figure 16, Sections C and D are the contents for students which have no made use of the MCI. We formulate two questions for the students that have made use of the MCI. These questions have four different concepts included. Question 1: How much has the use of the monitoring and control interface allowed you to assimilate clearly the following criteria? Question 2: How much has the use of the monitoring and control interface awakened your interest with respect to the following criteria? The criteria for both cases were: Process, control design, control law application and graphical interface interaction. In Sections C and D of Figure 16, the questions applied were as follows. Question 1: How much has the use of the interface enabled you to assimilate the following criteria? Question 2: How much has the use of monitoring and control interface sparked your interest in the following criteria?

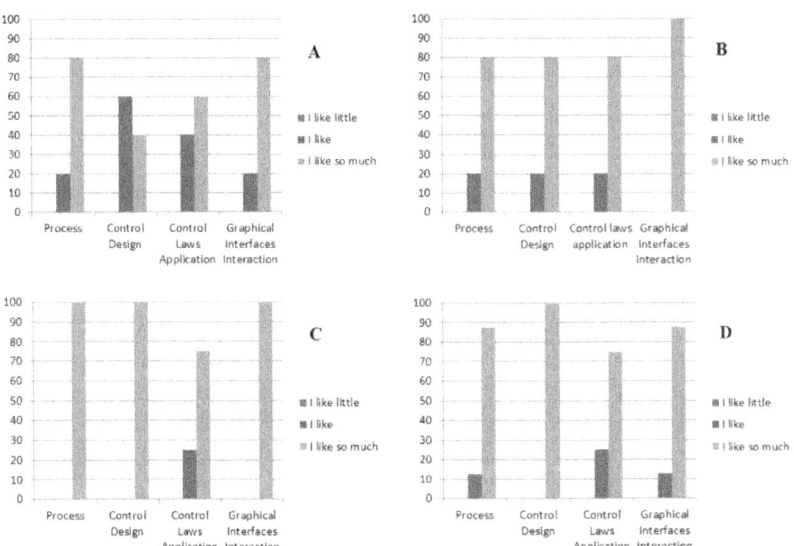

Figure 16. Statistics on the use of the monitoring and control interface (MCI).

Since the MCI has been in use, the interest of the students in heat exchanger control has increased, because, during the course period, the EES have acquired experience and practice in the operation of these devices. Furthermore, they have acquired knowledge in the control area and for how they can develop their control algorithms to design soft sensors or other control techniques.

CENIDET is the National Center for Research and Technological Development, which provides educational training support to technological universities. Students from 15 or 20 (approximately) technological universities of Mexico carry out professional practice in CENIDET, and in most of the cases, these students select CENIDET to develop their master studies. Some of the students that have had professional practice using the MCI and who are now working on their Master's thesis project are: Diego Alessis Carbot Rojas, "Actuator fault-tolerant control: Application in a heat transfer process"; Juan Pablo Castillo González, "Control design for a biodiesel reactor"; and Moisés Bulmaro Ramos Martínez, "Control design for an azeotropic process". Other success cases in the implementation of the MCI are students that decide to continue their Doctoral studies, students such as: Betty Yolanda López Zapa "Fault diagnosis applied in a biodiesel reactor"; Jarniel García Morales "Multiple-fuel control design for an internal combustion engine"; and Omar Hernández González "Observation and control of nonlinear process systems for energy recovery from waste heat". In addition, students have published articles in the control area using the heat exchanger device and the knowledge that the MCI provides: the most recent work done by a student is under review, however, in the Revista Mexicana de Ingeniería Química, this work is entitled Multiple Sensor Fault Diagnosis in a Heat Exchanger Using Sliding-Mode Observers Based on Super-Twisting Algorithm. Another example of the contribution made by the students is given in the work presented in [23], where the student, Betty Yolanda López-Zapata, was a collaborator. Furthermore, the MCI has provided support to develop some published works regarding heat exchanger devices and their control [24,25].

8. CONCLUSIONS

The monitoring control interface (MCI) is an excellent alternative to implement in a real heat exchange process in order to make the system fault tolerant. The MCI is capable of detecting and isolating failures using virtual sensors. These virtual sensors are accurate in outlet temperatures estimation. Furthermore, the MCI is an excellent option to strengthen the knowledge taught in the classroom. Subjects, such as classic control, nonlinear control and fault detection and isolation (via two different approaches, linear and nonlinear observers), are supported by this interactive educational workstation. The participation and the interest of the students in laboratory practices employing the heat exchanger pilot plant increased by more than 20% since the MCI was implemented, according to a survey. In a similar way, the lecturers were very interested in developing other algorithms to enrich the control area. For future work, subjects, such as fuzzy control, neural networks and system identification, could be added to the interactive educational workstation.

Acknowledgments

The authors would like to thank Tecnológico Nacional de México for the financial support to develop research on the control process (Project Number 5291.14-P), CENIDET (Centro Nacional de Investigación y Desarrollo Tecnológico), and for the support to develop and to apply the MCI.

Author Contributions

R.F. Escobar conceived and designed the experiments, performed the experiments and wrote the paper. M. Adam-Medina and C.D. García-Beltrán performed the experiments. V.H. Olivares-Peregrino and D. Juárez-Romero analyzed the data. G.V. Guerrero-Ramírez contributed reagents/materials/ analysis tools.

Appendix A

Nomenclature	
T	Temperature, (˚C)
U	Heat transfer coefficient, (W/m²K)
A	Heat transfer area, (m²)
C_p	Specific heat, (J/kg K)
Vl	Volume, (m³)
W_v	Volumetric flow, (cm³/min)
Greek letters	
ρ	Density, (kg/m³)
ζ	Error tolerance
θ	Observer gain
AT	Temperature increment
λ	Eigenvalue
ζ	Error tolerance
Subscripts	
c	Cold
h	Hot
i	Input
o	Output
ss	Steady state

Conflicts of Interest

The authors declare no conflict of interest.

REFERENCES

1. Arriandiaga, A.; Portillo, E.; Sánchez, J.A.; Cabanes, I.; Pombo, I. Virtual Sensors for On-Line Wheel Wear and Part Roughness Measurement in the Grinding Process. Sensors 2014, 14, 8756–8778.

2. Escobar, R.F.; Astorga-Zaragoza, C.M.; Hernández, J.A.; Juárez-Romero, D.; García-Beltrán, C.D. Sensor fault compensation via software sensors: Application in a heat pump's helical evaporator. Chem. Eng. Res. Des. 2014, doi:10.1016/j.cherd.2014.06.017, in press.

3. Ploennigs, J.; Ahmed, A.; Hensel, B.; Stack, P.; Menzel, K. Virtual sensors for estimation of energy consumption and thermal comfort in buildings with underfloor heating. Adv. Eng. Inform. 2011, 25, 688–698.

4. Astorga-Zaragoza, C.M.; Zavala-Río, A.; Alvarado, V.M.; Méndez, R.M.; Reyes-Reyes, J. Performance monitoring of heat exchangers via adaptive observers. Measurement 2007, 40, 392–405.

5. Yan, X.G.; Edwards, C. Nonlinear robust fault reconstruction and estimation using a sliding mode observer. Automatica 2007, 43, 1605–1614.

6. Tan, C.P.; Edwards, C. Sliding mode observers for detection and reconstruction of sensor faults. Automatica 2002, 38, 1815–1821.

7. Orani, N.; Pisano, A.; Usai, E. Fault diagnosis for the vertical three-tank system via high-order sliding-mode observation. J. Frankl. Inst. 2010, 347, 923–939.

8. Theilliol, D.; Noura, H.; Ponsart, J.C. Fault diagnosis and accommodation of a three-tank system based on analytical redundancy. ISA Trans. 2002, 41, 365–382.

9. Edwards, C.; Tan, C.P. Sensor fault tolerant control using sliding mode observers. Control Eng. Pract. 2006, 14, 897–908.

10. Fortuna, L.; Graziani, S.; Xibilia, M.G. Soft sensors for product quality monitoring in debutanizer distillation columns. Control Eng. Pract. 2005, 13, 499–508.

11. Escobar, R.F.; Astorga Zaragoza, C.M.; Hernández, J.A.; Adam Medina, M.; Guerrero Ramírez, G.V. Design and implementation of an observer-based soft sensor for a heat exchanger. Dyna 2011, 166, 89–97.

12. Douglas, F.; Soloway, E. Interface design: A neglected issue in educational software. In Proceedings of the SIGCHI/GI Conference on Human Factors in Computing Systems and Graphics Interface, Toronto, ON, Canada, 5–9 April 1987.

13. Cartaxo, J.M.; Fernandes, A.N. Educational software for heat exchanger equipment. Comput. App. Eng. Educ. 2010, 18, 193–199.

14. Machuca, F.; Urresta, O. Educational software for the teaching of the dynamics and control of shell and tube heat exchangers. Rev. Fac. Ing. Univ. Antioq. 2008, 44, 52–60.

15. Zueco, J. An educational laboratory virtual EES for encouraging the use of computer programming in thermal engineering problems. Comput. Appl. Eng. Educ. 2013, 21, 691–697.

16. Sieres, J.; Fernández-Seara, J. Simulation of compression refrigeration systems. Comput. Appl. Eng. Educ. 2006, 14, 188–197.

17. Cooper, D.; Dougherty, D. Enhancing process control education with the Control Station training simulator. Comput. Appl. Eng. Educ. 2000, 7, 203–212.

18. Zhi, Z.; Jiang, L.; Si, W.; Wei, Z. Thermodynamic Calculation and Numerical Simulation of Shell-and-tube Heat Exchanger. CNKI J. 2012, 1, doi:CNKI:SUN:HNYI.0.2012-01-008.

19. Granjo, J.F.; Rasteiro, M.G.; Gando-Ferreiraand, L.M.; Bernardo, F.P.; Carvalho, M.G.; Ferreira, A.G. A Virtual Platform to Teach Separation Processes. Comput. Appl. Eng. Educ. 2012, 20, 175–186.

20. Escobar, R.F.; Astorga-Zaragoza, C.M.; Téllez-Anguiano, A.C.; Juárez-Romero, D.; Hernández, J.A.; Guerrero-Ramírez, G.V. Sensor fault detection and isolation via high-gain observers: Application to a double-pipe heat exchanger. ISA Trans. 2011, 50, 480–486.

21. Rossiter, J.A. Predictive control—The basic algorithm. In Model Based Predictive Control a Practical Approach; University of Texas: Austin, TX, USA, 2004; pp. 32–58.

22. Malleswararao, Y.S.; Chidamaram, M. Nonlinear controllers for a heat exchanger. J. Process Control 1992, 2, 17–21.

23. Adam-Medina, M.; Escobar, R.F.; Juárez-Romero, D.; Guerrero-Ramírez, G.V.; López-Zapata, B. Fault Detection in a Heat Exchanger Using Second Order Sliding Mode Observers. Rev. Mex. Ing. Quím. 2013, 12, 327–336.

24. Zapata, B.L.; Escobar, R.F.; Medina, M.A.; Zaragoza, C.M. State variables estimation for a counter-flow double-pipe heat exchanger using Multi-linear Model. In Proceedings of the IEEE Electronics, Robotics and Automotive Mechanics Conference, Cuernavaca, Morelos, Brazil, 22–25 September 2009; pp. 349–354.

25. Zapata, B.L.; Medina, M.A.; Escobar, R.F.; Zaragoza, C.M. Diagnóstico de Fallas en Intercambiadores de Calor: Enfoque Multi-Modelos. Available online: http://amca.mx/memorias/amca2009/articulos/amca2009_90.pdf (access on 24 October 2014)

This page left intentionally blank.

INDEX